Dr. James Lee Johnson
1937–1977

KINETICS OF
COAL GASIFICATION

KINETICS OF COAL GASIFICATION

A Compilation of Research by
the late
Dr. James Lee Johnson

*Dedicated to his memory
by his associates at*
The Institute of Gas Technology

A WILEY-INTERSCIENCE PUBLICATION
JOHN WILEY & SONS
New York Chichester Brisbane Toronto

Copyright © 1979 by John Wiley & Sons, Inc.

All rights reserved. Published simultaneously in Canada.

Reproduction or translation of any part of this work beyond that permitted by Sections 107 or 108 of the 1976 United States Copyright Act without the permission of the copyright owner is unlawful. Requests for permission or further information should be addressed to the Permissions Department, John Wiley & Sons, Inc.

Library of Congress Cataloging in Publication Data:

Johnson, James Lee, 1937–1977.
    Kinetics of coal gasification.

    "A Wiley-Interscience publication."
    1. Coal gasification. 2. Chemical reaction, Rate of. I. Title.

TP759.J64 1979   665'.772   79-10439
ISBN 0-471-05575-1

Printed in the United States of America

10 9 8 7 6 5 4 3 2 1

In Memoriam . . .

Dr. James Lee Johnson

1937 – 1977

... *Dedicatory Statements*

Jim Johnson was my student, protégé, co-worker associate, and dear friend over a period of nineteen years. Very early in his career it was clear that he would become the world's leading coal kineticist and the worthy successor to Martin Elliott, Ev Gorin, Sabri Ergun, and others and, of course, his mentor, the late George von Fredersdorff. Tragically, fate stopped short this rapidly rising star, just as it did fifteen years ago at much the same point in the life of George von Fredersdorff. And like George, whose role he then filled, Jim was not only brilliant, but exceptionally gentle, kind, and thoughtful as well. Beyond that, he was in every way a credit to his profession and to the fuel-science fraternity to which he so uniquely contributed. We are fortunate that before his untimely death he completed his monumental "Fundamentals of Coal Gasification" chapter for the new edition of *The Chemistry of Coal Utilization*. As a part of this memorial volume, it will stand as a lasting testimonial to the technical and scientific accomplishments of Jim Johnson. But the monument to his humanity must remain in the hearts of those who knew him.

Dr. Henry R. Linden

President, Gas Research Institute
Chicago, Illinois
Former President, Institute of Gas Technology
Chicago, Illinois

On behalf of the National Research Council's Committee on the Chemistry of Coal Utilization, which was organized to supervise the publication of a supplementary volume to a book by that name, I want to pay tribute to Dr. James L. Johnson, the author of the chapter entitled, "Fundamentals of Coal Gasification." Dr. Johnson's work on the mechanism and kinetics of gasification reactions is outstanding and has received international recognition. This memorial volume will not only honor Dr. Johnson's memory, but also, as a collection of his publications, will remind us of his many contributions to science and technology.

George R. Hill

*Chairman, Committee on the
  Chemistry of Coal Utilization
National Research Council
Washington, D.C.*

I knew Jim Johnson as a graduate student and was closely associated with him at various times throughout his all too short professional career. He was an outstanding scientist and technologist and a fine human being, who made major contributions to our knowledge of the mechanism and kinetics of gasification reactions. In this memorial volume, but a small tribute to Jim and his accomplishments, the collection of his publications attests to the capability of the man and is a richly deserved recognition of the value of his work.

Martin A. Elliott

*Technical Editor
Second Supplementary Volume of
  the Chemistry of Coal Utilization*

Jim Johnson was one of the best. He will be much missed by those who know him and others who found his work helpful. Jim's principal activity was his study of the reaction kinetics for the gasification and hydrogasification of coal. His greatest contribution to this increasingly important field of knowledge was his development of a mathematical model for these reactions, which model had general applicability to all such processes. He was very innovative in his approaches. He was helpful to others in the field and very cooperative. Most recently, he wrote a chapter for the second supplementary volume of *The Chemistry of Coal Utilization* on the fundamentals of coal gasification. This chapter was especially well organized and definitive in its presentation of the literature. Unknown to him at the time, it was to become the crowning glory of his distinguished career.

R. Tracy Eddinger

Technical/Engineering Manager
COGAS Development Company
Princeton, New Jersey

One of the benefits of working on committees is the intellectual stimulation of association with highly qualified persons, dedicated to the advancement of ideas —Jim was a fine example. He worked hard to achieve a result that would credit not only himself, but the whole effort of which he was a part. As short as our relationship was in its duration, it was long in the satisfaction of knowing that a job worth doing had been done well.

L. S. Galstaun

Manager, Process Design
Bechtel International
Houston, Texas

Through his extensive laboratory measurements and mathematical modeling, Jim Johnson made unique and lasting contributions to the development of technology for coal hydrogasification. His work at the interface between fundamental knowledge and practical need constitutes a significant step toward the establishment of a predictive understanding of the complex hydrogasification phenomena.

Jack B. Howard

Professor of Chemical Engineering
Massachusetts Institute of Technology
Cambridge, Massachusetts

It is a sad duty to record the premature death of James L. Johnson, author of the chapter on "Fundamentals of Gasification." It is particularly sad for those who had the opportunity to work with him and to know and appreciate his dedication to the subject. This chapter in the third edition of Lowry's *Handbook* will stand as a memorial to the promise of his future contributions that must now remain unfulfilled.

Eric H. Reichl

Former President
Conoco Coal Development Company
Stamford, Connecticut

As this nation moves to greater dependence on the conversion of coal to a clean energy source, it will become more appreciative of the brilliance and dedication of Dr. James Johnson and his accomplishments in bringing an understanding to the chemical reactivity and complexities of this abundant solid. His contributions to coal science in this area are unsurpassed and will provide guidance to other investigators for years to come.

Dr. Robert C. Weast

Vice President, Research
Consolidated Natural Gas Service Company
Cleveland, Ohio

Contents

1 Fundamentals of Coal Gasification *1*

2 Kinetics of Coal Gasification in Hydrogen during Initial Reaction Stages *179*

3 Gasification of Montana Lignite in Hydrogen and Helium during Initial Reaction Stages *205*

4 Relationship Between Gasification Reactivities of Coal Char and Physical and Chemical Properties of Coal and Coal Char *237*

5 The Use of Catalysts in Coal Gasification *261*

6 Kinetics of Bituminous Coal Char Gasification with Gases Containing Steam and Hydrogen *283*

Index *317*

KINETICS OF
COAL GASIFICATION

1 Fundamentals of Coal Gasification

INTRODUCTION

The subject of coal gasification covers the conversion of coals to light gases, condensable liquids and tars, and solid products in the presence of reactive gaseous atmospheres. Thus gasification is distinct from other forms of coal conversion such as pyrolysis or carbonization, which occur in inert-gas atmospheres, or liquefaction, which occurs in liquid medium. Research efforts to develop the technology for converting coals into gases have been greatly accelerated since the early 1960s, particularly in the United States, due to projected shortages of other economically competitive fuels for use as energy sources. These expanded efforts have been focused on reaction systems relevant to the conversion of coals into relatively low-energy gases suitable for power generation and other industrial uses, as well as into high-energy gases that can be economically transported over long distances and used as a substitute for natural gas.

The objective of the present work is to discuss fundamental aspects of coal gasification that have been developed in the many investigations conducted during the past 15-20 years in the areas of thermodynamics, physics, and chemical kinetics. Because this subject area is very broad, selection of the specific topics for discussion has necessarily been subjective and has depended on individual interests, particularly those concerning physical processes involved in coal-gasification systems, for which a vast literature exists. Discussion of the thermodynamics and kinetics of coal gasification, however, extends to coal-char gasification reactions as well as to the initial coal-gasification processes, in which volatile matter is evolved and coal char is formed in reactive atmospheres. Elucidation of the kinetic behavior during the initial coal-gasification stages is particularly important to understanding the conversion of coal to methane at elevated pressures.

Published by permission of the National Academy of Sciences, Washington, D. C., the copyright holder. To appear in *The Chemistry of Coal Utilization, Supplementary Volume II*, Martin Elliot, Ed., Wiley, New York, forthcoming.

2 FUNDAMENTALS OF COAL GASIFICATION

THERMODYNAMICS OF GASIFICATION

Thermodynamic considerations related to the gasification of coals and coal chars are important to the theoretical evaluation of limiting performance characteristics for individual process concepts and to the practical design of experimental or commercial reactor systems. For these purposes, thermochemical properties of coal-gasification systems must be known or estimated to define heat effects and equilibrium or pseudoequilibrium behavior. Measurement of the thermochemical properties of coals and coal chars, however, is very difficult because of the complex and heterogeneous nature of such materials in comparison to graphite. Furthermore, there is uncertainty as to the accuracy and even the meaning of much of the available experimental information.

The following discussion focuses on equilibrium activity coefficients, specific heats, and heats of carbonization and hydrogenation reactions as they apply to coal- and coal-char-gasification systems.

Equilibrium Effects

To determine the equilibrium characteristics of coal- or coal-char-gasification systems, it is necessary to define the standard free energies of formation of the reactants and products. (The standard heats and entropies of formation can also be used.) For the major gaseous species involved in such systems, this information is available in a variety of tabulations and correlations. However, due to a lack of accurate experimental information, relatively crude techniques have been used to estimate free energies of formation or equivalent parameters for coals and coal chars.

One technique to estimate coal thermodynamic properties has been to extrapolate or interpolate from the properties of pure, solid aromatic compounds [1]. It is not easy, however, to evaluate the uncertainties that result from using this technique, not even taking into consideration the complex heterogeneous character of coal as well as its amorphous nature. The use of the estimated free energies of formation of coal itself in thermodynamic analyses of coal-gasification systems is perhaps most applicable for estimating overall potential performance characteristics of gasification concepts by accounting for second-law constraints.

Some investigators have interpreted experimental yields obtained during the gasification of coals- or coal-chars in terms of pseudoequilibrium constants, or equivalently, by assigning "activity" coefficients to the reacting carbon. This has been done when the investigator is interested in the methane yields resulting from gasification with hydrogen or with synthesis gas mixtures, according to the following reaction:

$$C + 2H_2 \Longleftrightarrow CH_4 \qquad [1]$$

The equilibrium constant for this reaction is generally expressed by the following equation:

$$K_{CH_4} = \frac{P_{CH_4}}{a_C P_{H_2}^2} \quad \text{or} \quad K'_{CH_4} = a_C K_{CH_4} = \frac{P_{CH_4}}{P_{H_2}^2} \quad [1]$$

where

P_{H_2}, P_{CH_4} = partial pressure of H$_2$ and CH$_4$, respectively (in atm)

a_C = "activity" of a particular carbon referred to graphite (for graphite $a_C = 1$)

Wen, Abraham, et al. [2] evaluated experimental values of $P_{CH_4}/P_{H_2}^2$ obtained at apparent pseudoequilibrium conditions during the gasification, with hydrogen, of low-temperature bituminous coal char. Gasification occurred at elevated pressure in a circulating gas system at 705°C. Values of K'_{CH_4}, reported as a function of total carbon gasified, decreased from about 0.8 atm^{-1} at zero carbon conversion to about 0.12 atm^{-1} at carbon-conversion fractions approaching unity. Based on Equation 1, this corresponds to a range of apparent carbon activity, a_C, from 7 to 1.

Similarly, Squires [3] compiled plots of values of $P_{CH_4}/P_{H_2}^2$ from data obtained in integral steam-oxygen gasification of coals and coal chars and showed an apparent systematic trend in which values of a_C are generally greater than unity but decrease with increasing pressure, approaching unity at total pressures above 50 atm. Although apparent values of a are relatively independent of temperature at 760-900°C, values as high as $a_C = 20$ are indicated for data obtained at 1 atm. of total pressure.

There are several possible explanations for the experimental values of a_C greater than unity obtained in coal- and coal-char-gasification systems. Squires [3] suggests that amorphous carbons have excessive free energies compared to graphites. He also suggests that, in gases containing steam as well as hydrogen, the steam activates carbons for gasification with hydrogen. Blackwood and McCarthy [4], however, have argued that for gasification of carbons in gases containing both steam and hydrogen it is not valid to interpret apparent carbon activities solely on the basis of the reaction given by Reaction 1, since methane can be formed by other reactions involving both steam and hydrogen. As indicated by Blackwood and McCarthy, for integral fluidized-bed gasification of brown coal with hydrogen at elevated pressures, Birch, Hall, et al. [5] found that $P_{CH_4}/P_{H_2}^2$ ratios in product gases never exceed equilibrium values where $a_C = 1$. Blackwood and McCarthy also noted that the equilibrium measurements of hydrogen-methane mixtures made by Browning and Emmett [6] using a carbon formed from ion carbide decomposition corresponded closely to equilibrium values obtained for graphite gasification in hydrogen ($a_C = 1$) over a wide range of temperatures. Such evidence prompted Blackwood and McCarthy to suggest that coal-char-gasification systems are generally bounded by equilibria in which carbon activities are approximately the same as for graphite.

Johnson [7] has also shown that results such as those interpreted by Squires to reflect values of a_C greater than unity can be reasonably simulated with a kinetic model accounting for the reactions of steam as well as hydrogen with carbon. Significantly, Johnson's kinetic expressions all involve equilibrium terms in which coal-char activities are assumed to be the same as those of graphite. Such analyses emphasize the possible dangers of interpreting thermodynamic activities based on the characteristics of a reaction subsystem when the total reaction system is not in equilibrium.

Another assumption sometimes made in estimating the thermodynamic characteristics of coal chars is that the higher heats of formation of such materials with respect to graphite are directly related to higher free energies of formation. This assumption can have significant consequences. If, for example, an amorphous carbon having a heat of formation of 2.6 kcal/mol is assumed to have an increased free energy of formation of 2.6 kcal/mol relative to graphite (independent of temperature), then values of a_C close to 3.4 would result at 800°C. This approach does not recognize, however, is that amorphous carbons would also be expected to have higher entropies than graphite, which could either partially or wholly compensate for the higher heats of formation.

For experimental systems in which coal-devolatilization reactions are occurring in addition to the subsequent gasification of coal char, apparent high values of a_C can also result. This occurs because coal-decomposition reactions are generally irreversible, and the methane yields obtained are a result of relative kinetic factors rather than equilibrium or pseudoequilibrium characteristics, particularly in the presence of hydrogen at elevated pressures. This is discussed in a later section of this chapter concerned with initial-stage coal-gasification kinetics. Such a consideration, however, suggests that apparent equilibrium constants, such as those obtained by Wen, Abraham et al. [2] using a low-temperature coal char containing significant volatile matter, may reflect kinetic behavior that depends on the specific reaction conditions employed and should thus be applied or extrapolated with caution. In any case, it is apparent that at this time there is no general correlation defining effective char-carbon activity coefficients in terms of temperature, pressure, gas composition, the reaction that is occurring, and any known property of the solid phase.

The gas composition obtained under equilibrium conditions for the gasification of coal chars can be at least qualitatively estimated by assuming that the solid carbon has the same free energy of formation as that of graphite. Based on the preceding discussion, it is quite possible that such an assumption may be quantitatively applicable to a wide range of conditions relevant to coal-char-gasification systems. Such equilibrium calculations have been made by a number of authors, including von Fredersdorff and Elliott [8].

For the graphite-hydrogen-methane system, equilibrium gas composition for Reaction 1 can be expressed only as a function of temperature and total pressure. Equilibrium methane mole fractions for this system (shown in Figure 1) are based on computations using thermodynamic data from the JANAF Thermochemical Tables [9], in which gas-phase species are assumed to be ideal gases. As indicated, equilibrium methane mole fractions decrease with increasing temperature and decreasing pressure.

For the graphite-carbon dioxide-carbon monoxide system, equilibrium gas compositions for the reaction

$$C + CO_2 \Longleftrightarrow 2CO \qquad [2]$$

are also a function only of temperature and total pressure. Equilibrium carbon monoxide mole fractions for this system (shown in Figure 2) increase with in-

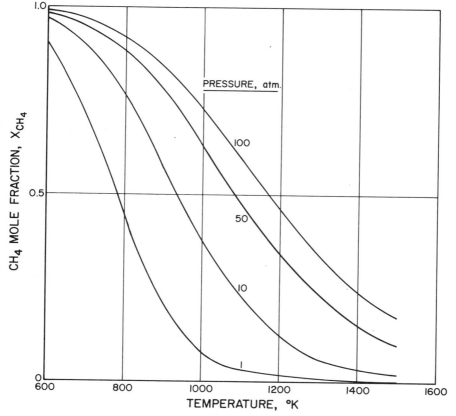

Figure 1 Equilibrium characteristics for $C(graphite)$-H_2CH_4 system.

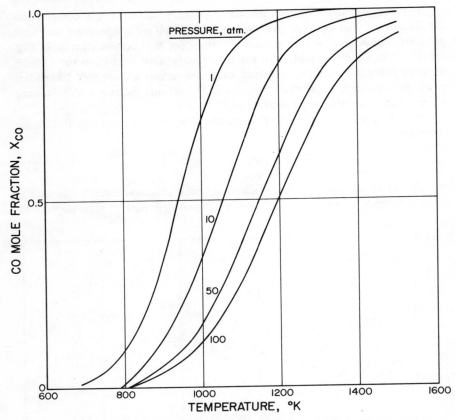

Figure 2 Equilibrium characteristics for C(graphite)–CO–CO$_2$ system.

creasing temperature and decreasing pressure, a trend opposite to that exhibited for methane mole fraction in the graphite-hydrogen-methane system.

In a C-H-O gasification system in which the components carbon, hydrogen, and oxygen are all present, it is usually assumed that the following species exist in the gas phase: CO, CO$_2$, H$_2$, H$_2$O, and CH$_4$. The equilibrium composition for this system can be expressed as a function of temperature, total pressure, and the hydrogen:oxygen ratio (atom:atom) in the gas phase. For purposes of calculation, it can be assumed that the equilibrium reactions are comprised of those given by Reactions 1 and 2, as well as the water-gas shift reaction

$$CO + H_2O \rightleftharpoons CO_2 + H_2 \qquad [3]$$

This selection of reactions is, of course, arbitrary since a variety of equivalent combinations of reactions could be chosen for equilibrium calculations. An example of equilibrium trends in this more complex system is shown in Figure 3 for an hydrogen:oxygen ratio of unity. These results are based on calculations made by Baron, Porter, et al. [10], who also used the JANAF Thermochemical Tables and assumed ideal-gas properties. The trends exhibited in Figure 3 for hydrogen:oxygen ratio of 1 can be more generally characterized as shown in Table 1.

The arrows in Table 1 indicate whether a variable increases (↑), decreases (↓), or initially increases, reaches a maximum, and then decreases (→). For example, the carbon monoxide mole fraction increases with increasing temperature but decreases with increasing pressure and increasing hydrogen:oxygen

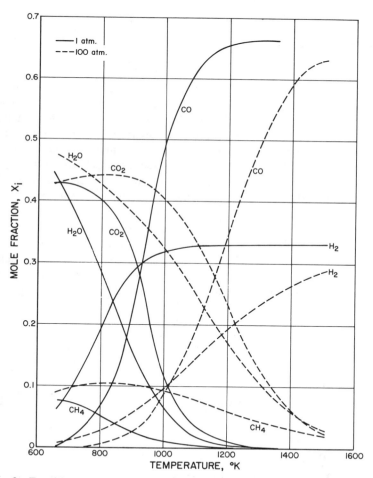

Figure 3 Equilibrium characteristics for C-H-O system (hydrogen:oxygen = 1 g-atom/g-atom).

TABLE 1. EQUILIBRIUM TRENDS FOR C-H-O SYSTEM (HYDROGEN:OXYGEN RATIO = 1 g-atom/g-atom)

Mole Fraction, X_i	Temperature ↑	Pressure ↑	Hydrogen:Oxygen Ratio ↑
X_{CO}	↑	↓	↓
X_{CO_2}	→	↑	↓
X_{H_2}	↑	↓	↑
X_{H_2O}	↓	↑	→
X_{CH_4}	→	↑	↑
$\left[\dfrac{X_{CH_4}}{X_{CH_4} + X_{CO} + X_{CO_2}} \right]$	→^{a}, ↓^{b}	↑	↑

^{a} Hydrogen:oxygen < 2.
^{b} Hydrogen:oxygen ⩾ 2.

ratio. The carbon dioxide mole fraction reaches a maximum value with increasing temperature, increases with increasing pressure, and decreases with increasing hydrogen:oxygen ratio. The relative amount of gas-phase carbon as methane, given by $X_{CH_4}/(X_{CO} + X_{CO_2} + X_{CH_4})$, increases with both increasing pressure and increasing hydrogen:oxygen ratio. The effect of increasing the temperature on this variable depends on the hydrogen:oxygen ratio. Relative methane yields have a maximum value for hydrogen:oxygen values less than 2, although they decrease continuously with increasing temperature for hydrogen:oxygen values equal to or greater than 2.

The equilibrium characteristics shown in Figures 1-3 are all constructed on the assumption that the solid carbon phase has the same thermodynamic properties as does graphite. For the case in which the solid carbon has an activity greater than that of graphite, that is, a greater free energy of formation, the equilibrium constants for Reactions 1 and 2 can be more generally represented by

$$K_{CH_4} = \frac{X_{CH_4}}{X_{H_2}^2 (P \cdot a_C)} \qquad [2]$$

and

$$K_{CO} = \frac{X_{CO}^2}{X_{CO_2}} \left(\frac{P}{a_C} \right) \qquad [3]$$

where

P = total pressure (in atm)
X_i = mole fraction of i species
K_{CH_4}, K_{CO} = equilibrium constants for Reactions 1 and 3, respectively

In Equations 2 and 3 the equilibrium constants are the same as those for the gasification of graphite, where $a_C = 1$, by definition. For values of a_C other than unity, the equilibrium characteristic for the C-H$_2$-CH$_4$ and the C-CO-CO$_2$ systems can still be computed from equilibrium characteristics where $a_C = 1$ by defining

$$P' = P \cdot a_C \qquad [4]$$

$$P'' = \frac{P}{a_C} \qquad [5]$$

If, for example, $P = 10$ atm and $a_C = 5$, then the equilibrium methane mole fraction can be determined from Figure 2 by using an effective pressure of $P' = 10 \times 5 = 50$ atm.

Increasing the value of a_C generally leads both to increased methane and carbon monoxide mole fractions in each of the simple systems considered. The situation in the C-H-O system is more complex, and the generalized effects of increasing a_C have not been evaluated in this analysis. However, an example has been computed for the case in which $P = 100$ atm, $T = 1100°K$, the hydrogen:oxygen ratio = 1, and the a_C has values of 1, 2, and 4. The results of the calculation are given in Table 2. They indicate that under these conditions, not only do methane and carbon monoxide mole fractions individually increase with increasing a_C, but that the relative amount of carbon in the gas present as methane, $X_{CH_4}/(X_{CO} + X_{CO_2} + X_{CH_4})$, also increases with increasing a_C. In addition, the total carbon:oxygen ratio in the gas $(X_{CO} + X_{CO_2} + X_{CH_4})/(X_{CO} + 2X_{CO_2} + X_{H_2O})$ increases with increasing a, which indicates greater gasification yields.

TABLE 2. EFFECT OF VARYING a_C

	a_C		
Mole Fraction	1	2	4
X_{CO}	0.19	0.27	0.35
X_{CO_2}	0.33	0.32	0.29
X_{H_2}	0.15	0.13	0.11
X_{H_2O}	0.25	0.16	0.09
X_{CH_4}	0.08	0.12	0.16
Total	1.00	1.00	1.00
$X_{CH_4}/(X_{CO} + X_{CO_2} + X_{CH_4})$	0.13	0.17	0.20
$(X_{CO} + X_{CO_2} + X_{CH_4})/(X_{CO} + 2X_{CO_2} + X_{CH_4})$	0.54	0.67	0.79

Heat Effects

The effects of heat on coal-gasification systems can be expressed in terms of the sensible heats of the reacting substances and the heats of reaction. Specific heats of coals and coal chars, necessary to evaluate the sensible heat contributions, have been measured using a variety of experimental techniques [11-23]. For the most part, studies with coals have been conducted at temperatures below about 400°C to avoid the complications that result from the heats of coal-decomposition reactions. Although studies have also been made over a higher temperature range using coals and coal chars containing volatile matter, [11, 23-27], measured heat effects under such conditions are due to both the sensible heat effects and the heats of reaction. In such cases, either apparent or average pyroheat capacities have been reported, or some attempt has been made to distinguish between sensible heat and reaction-heat contributions to total heat effects.

A novel and apparently useful correlation for estimating actual specific heats of coal and its solid-conversion products has been proposed by Kirov [28] based on an analysis of selected literature data. In his correlation the specific heat of a coal or coal char is assumed to be the weighted summation of the specific heats of the major coal components, that is

$$C_p = X_v C_v + X_a C_a + X_w C_w + X_C C_C \qquad [6]$$

where

X_v, X_a, X_w, X_C = mass fraction of the potential volatile matter, ash, water, and fixed carbon, respectively

C_v, C_a, C_w, C_C = specific heats of potential volatile matter, ash, water, and fixed carbon, respectively (in cal/g-°C)

C_p = average specific heat of coal or coal char (in cal/g-°C)

The specific heats are represented as functions of temperature with the exception of C_w, which is assumed to be constant at 1 cal/g-°C. Correlations proposed for the ash and fixed carbon fractions are

$$C_a = 0.18 + 1.4 \times 10^{-4}\ t' \qquad [7]$$

$$C_C = 0.165 + 6.8 \times 10^{-4}\ t' - 4.2 \times 10^{-7} (t')^2 \qquad [8]$$

where t' is temperature (in °C). Kirov assumed that the total potential volatile matter of a coal or coal char is divided into two types: primary volatile matter (type 1), which is evolved at relatively low temperatures, and secondary volatile matter (type 2), which is evolved at higher temperatures. The specific heat for potential primary volatile matter, C_{v_1}, was approximated by the specific heat reported by Cragoe [29] for a low-temperature tar of specific gravity 1.04 in the liquid phase:

$$C_{v_1} = 0.395 + 8.1 \times 10^{-4} \, t' \qquad [9]$$

The specific heat of the potential secondary volatile matter was represented by

$$C_{v_2} = 0.71 + 6.1 \times 10^{-4} \, t' \qquad [10]$$

Equation 10 was derived from Clendenin, Barclay, et al.'s [30] formulation of the specific heats of anthracites as a function of temperature and volatile matter, extrapolated for the case of 100% volatile matter.

Kirov defined the relative amounts of potential primary and secondary volatile matter in a coal solid by assuming that for the total volatile matter mass fraction (V_T) of 0.1 or less than an maf basis, all volatile matter is secondary volatile matter. For total volatile matter mass fractions greater than 0.1 (maf), the secondary volatile matter mass fraction, X_{v_2}, is assumed to be equal to 0.1 $(1 - X_a - X_w)$, with the remaining volatile matter existing as primary volatile matter (X_{v_1}). These assumptions are described by the following expressions:

$$V_T = \frac{X_{v_1} + X_{v_2}}{1 - X_a - X_w} \qquad [11]$$

for

$$V_T \leqslant 0.1, X_{v_1} = 0 \quad \text{and} \quad X_{v_2} = (1 - X_a - X_w) V_T \qquad [12]$$

for

$$V_T > 0.1, X_{v_1} = (V_T - 0.1)(1 - X_a - X_w),$$
$$X_{v_2} = 0.1 (1 - X_a - X_w) \qquad [13]$$

Kirov compared the predictions of his total correlation with experimental data available in 1965 for mean specific heats of different coals over the temperature range of 20–100°C. Here, mean specific heat, $\overline{C}_p$, is defined as

$$\overline{C}_p = \frac{\int_{t_0'}^{t_f'} C_p \, dt'}{t_f' - t_0'} \qquad [14]$$

where

t_0' = initial coal temperature
t_f' = final coal temperature

12 FUNDAMENTALS OF COAL GASIFICATION

For coals ranging approximately 0–50% volatile matter, agreement with the correlation was obtained to within about ±7%.

In a recent study at the Institute of Gas Technology (IGT) [31], Kirov's correlation was compared to the mean specific heats of coals and coal chars at temperatures of up to 300°C, and good agreement was found for coals ranging about 0–50% volatile matter. In this IGT study mean specific heats of low-volatility coal char at 25°–600°C were also found to be accurately predicted by Kirov's correlation. Since Kirov's correlation accounts only for specific heats of ungasified components in coals or coal chars, there is little basis to evaluate its applicability at higher temperatures, where significant thermal decomposition can occur, leading to reaction-heat effects.

Although Kirov suggests that net reaction-heat effects (with the exception of water vaporization) are negligible under carbonization conditions, some measurements of the total heats of coking have indicated that estimates based on Kirov's correlation are low by more than 15% [25]. Assuming that specific heats of carbonizing coal solids are reasonably described by his correlation, this would suggest that net reaction-heat effects during carbonization are somewhat endothermic. There is, however, no real agreement in the literature concerning the endothermicity or exothermicity of reaction-heat effects during coking, apparently because of the complications in distinguishing between relatively small reaction heat effects and sensible heat requirements.

Net heat effects during the carbonization of coals and coal chars of varying volatile content were reported by Lee [26] based on results of an experimental calorimeter study. With the technique employed, measurements were made of total heat requirements for the transition of feed solids at 21°C to carbonized solid residue and gasified products at calorimeter temperature, t'. Results were reported in terms of

$$\overline{C}_{pm} \equiv \frac{Q}{t' - 21} \qquad [15]$$

where

$\overline{C}_{pm}$ = proheat capacity (in cal/g-°C)
Q = total heat requirement (in cal/g)
t' = temperature (in °C)

Values of $\overline{C}_{pm}$ obtained at final temperatures of 300–800°C for feed solids having a volatile matter content of approximately 3–43 wt% (dry) were correlated with the following expression (modified to adjust units):

$$C_{pm} = 0.174 + 1.98 \times 10^{-4} \, t' + (0.33 + 5.49 \times 10^{-4} \, t') \cdot V \qquad [16]$$

where

t' = final temperature in °C
V = volatile matter of feed coal or coal char, weight fraction (dry)

In the development of Equation 16, the effect of the ash content in the feed solids used, which ranged 9-25% (dry), was not considered, nor was an adjustment made for the approximately 1% moisture in the feed solids. In effect, Equation 16 reflects the total heat, given by $Q = \overline{C}_{pm}(t' - 21)$, that is required to heat the feed solids from 21°C to t', plus the heat required for water vaporization and any reactions that occur in the system, including thermal decomposition, polymerization, and secondary gas-phase reactions. Since total solids conversion is not a parameter in Equation 16, it is important to note that measurements were made under conditions in which solids were brought to calorimeter temperature in about 5-10 min and maintained at temperature for about 90 min. This is significant because product yields could vary for other heat-up rates and residence times, leading to different heat effects.

A plot of sensible heat effects predicted by Kirov's correlations [28] and of total heat effects predicted by Lee's correlation [26] for the carbonization of coals or coal chars with 10%, 30%, and 50% volatile matter (maf) is shown in Figure 4. Values of $\overline{C}_p$ were computed from Kirov's correlation under the assumption that $X_a = 0.17$ and $X_w = 0.01$, which is consistent with the average concentrations of these components in the coal chars used by Lee to develop Equation 16. Figure 4 shows that Lee's correlation for coals of 30% and 50% volatile matter generally predicts greater heat requirements than does Kirov's correlation. If these differences are attributed to endothermic heats of reaction during carbonization, then at 800°C and 30-50% feed volatile matter, reaction-heat values will vary from 65 to 125 cal/g-feed coal solids (maf). Although the basis used to compute these values is subject to considerable uncertainty, these heat effects are relatively small compared to the heats of reaction of graphite with various gases, as shown in the table that follows, where thermodynamic data are from the JANAF tables [9].

Reaction	Heat of Reaction at 800°C (cal/g Carbon Gasified)
$C + O_2 \longrightarrow CO_2$	-7857
$C + 2H_2 \longrightarrow CH_4$	-1798
$C + H_2O \longrightarrow CO + H_2$	2705
$C + CO_2 \longrightarrow 2CO_2$	3382

Mahajan, Tomita, et al. [32] have also reported carbonization heats of a variety of coals, based on differential scanning calorimetry (DSC) studies. Fig-

14 FUNDAMENTALS OF COAL GASIFICATION

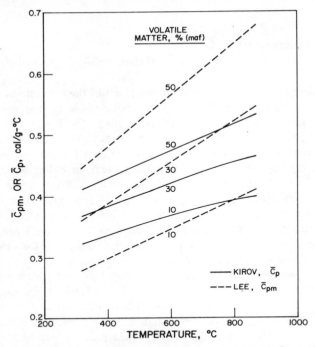

Figure 4 Comparison of correlations for carbonization heat effects.

ure 5 shows total heats of reaction determined as a function of feed volatile matter for coals heated to 580°C at 10°C/min, in helium at 55 atm. With this treatment, incomplete devolatilization was observed for all coals tested by Mahajan and colleagues, and the ratio of total weight loss to initial volatile matter ranged about 0.5–0.75. Also shown in Figure 5 are carbonization heats computed from the differences in heat requirements predicted by Kirov's and Lee's correlations at 580°C for concentrations of 30% and 50% feed volatile matter (maf). As is apparent, endothermic values obtained from Kirov's and Lee's correlations are consistent only with the largest endothermic heats of reaction reported by Mahajan and colleagues. The general scatter in the data of the Mahajan group probably emphasizes the complications and uncertainties that exist in obtaining accurate and well-characterized reaction-heat data in coal carbonization and gasification systems. The proper distinction between sensible heat and reaction-heat effects represents only one of the many difficulties involved in quantitatively interpreting results of calorimetry studies as well as DSC or differential thermal analysis (DTA) studies.

Efforts have also been made to evaluate heat effects during the gasification of coals and coal chars in hydrogen at elevated pressures. In such a reactive gaseous environment, total heat measurements comprise sensible heat effects,

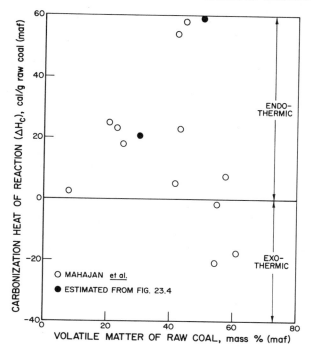

Figure 5   Effect of coal rank on carbonization heat of reaction.

heats of reaction due to thermal pyrolysis, heats of reaction due to direct gas-coal or coal-char interactions, and heats of reaction of secondary gas-phase interactions. Lee, Feldkirchner, et al. [33] have reported reaction-heat measurements for bituminous coal and coal chars gasified in hydrogen, which were based on calorimetry studies at 68 atm and temperatures of 705°C and 816°C. This study employed the same experimental procedure used to evaluate carbonization heat effects [26], with reaction products retained in the calorimeter for periods of up to 300 min. In his study with hydrogen, Lee also measured total solid conversion and computed net exothermic heats of reaction per mass of coal gasified, $-\Delta H_r$, by adjusting total heat effects according to

$$-\Delta H_r = \frac{-Q + \overline{C}_{pm}(t' - 21)}{\Delta m} \quad [17]$$

where

Q = total heat input (in cal/g feed solids)
$\overline{C}_{pm}$ = pyroheat capacity (calculated from Equation 16) (in cal/g-°C feed solids)
Δm = fraction of feed coal gasified (in g/g feed solids)

16 FUNDAMENTALS OF COAL GASIFICATION

One interpretation of Equation 17 is that the kinetics and thermal characteristics of carbonization depend only on temperature, even in a reactive atmosphere where additional reaction processes occur. Using this simplifying assumption, it can be seen that $-\Delta H_r \cdot \Delta m$ reflects the net heat of reaction of the additional reaction processes that occur in the hydrogen environment, such as carbon gasification to form methane ($C + 2H_2 \longrightarrow CH_4$) and the secondary hydrogenation of gasified volatile matter.

Figure 6 shows values of $-\Delta H_r$ obtained by Lee [33] as a function of the fraction of coal solids reacted. Although temperature does not appear to have a significant effect on values of $-\Delta H_r$, the results do indicate a trend in which $-\Delta H_r$ decreases with increasing volatile matter in the feed solids. Although Lee has suggested that this trend might be due to the effects of increased endothermic pyrolysis with the higher-volatility materials, his conclusion appears to conflict with the use of pyroheat capacity in adjusting total heat measurements, which ideally should compensate for pyrolysis heat.effects.

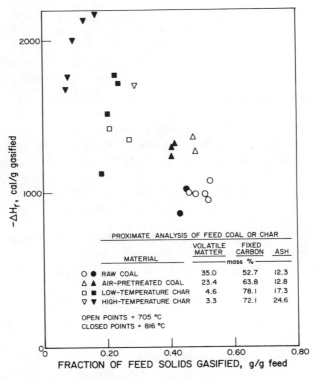

Figure 6   Effect of solid conversion on hydrogenation heat reaction.

Lee's results can also be interpreted by referring net heat effects only to the fixed carbon gasified in Lee's calorimeter experiments and not to the total solid conversion, including volatile matter gasified. This approach assumes that "fixed carbon" and "volatile matter" are independent components in coal solids, a simplifying assumption that is not valid over a wide range of conditions. Let $-\Delta H_r' = \Delta H_r \cdot \Delta m$ = net exothermic heat of reaction in cal/g-feed solids. Also assume that the net heat of reaction, $-\Delta H_r'$, is the sum of the heat of reaction of fixed carbon hydrogenation plus the heat of hydrogenation of gasified volatile matter:

$$-\Delta H_r' = h_f \Delta F + h_v \Delta V \qquad [18]$$

or

$$-\Delta H_r' = h_f \Delta F + h(\Delta m - \Delta F) \qquad [19]$$

where

ΔF = fixed carbon gasified (in g/g feed solids)
$\Delta V = \Delta m - \Delta F$ = volatile matter gasified (in g/g feed solids)
h_f, h_v = average heats of reaction of fixed carbon gasification and of the gas-phase hydrogenation of volatile matter, respectively (in cal/g)

In addition, it can be expected that under the conditions of Lee's experiments, with residence times of up to 300 min, most of the pyrolysis hydrocarbons, other than heavy tars, would be hydrogenated to methane [34]. Equation 21 can be rearranged to yield

$$-\Delta H_r'' = \frac{-\Delta H_r'}{\Delta F} = \frac{-\Delta H_r \cdot \Delta m}{\Delta F} = h_f + h_v \left(\frac{\Delta m}{\Delta F} - 1 \right) \qquad [20]$$

Thus a plot of $-\Delta H_r''$ versus $(\Delta m / \Delta F) - 1$ should yield a straight line with a y-axis intercept equal to h_f and a slope equal to h_v. Such a plot, shown in Figure 7, is based on calculations made with Lee's data, in which ΔF values were computed from reported proximate analyses of solid feed and residues.

In plotting the line shown in Figure 7, it was arbitrarily assumed that for zero volatile matter gasified ($\Delta m = \Delta F$), h_f is equal to the heat of the hydrogen gasification of graphite. Based on this assumption, a slope of h_v = 650 cal/g was obtained. At best, this value of h_v is applicable only to the bituminous coal and coal chars used by Lee, since the distribution of volatile products would vary for different coal types. In addition, since not all of the volatile matter can be hydrogenated, h_v does not represent an average heat of reaction per mass of vola-

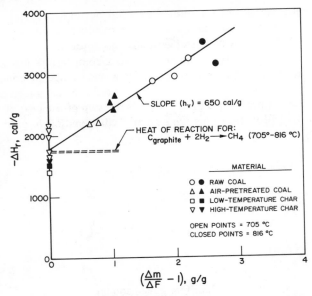

Figure 7 Correlation of hydrogenation heats of reaction.

tile matter actually hydrogenated, but rather the heat of reaction per mass of total volatile matter gasified.

A more detailed representation of h_v is

$$h_v = \sum_i X_i h_{v_i} \qquad [21]$$

where

X_i = mass fraction of species i in gasified volatiles (in g/g volatiles gasified)
h_{v_i} = heat of hydrogenation for conversion of species i to methane (in cal/g-species i)

Equation 21 shows that the value of h_v depends on the distribution of the species gasified during devolatization. For example, it can be assumed that the raw bituminous coal used by Lee (see Figure 6 for proximate analysis) has the distribution of volatiles shown in Table 3. Based on this simplified distribution of volatile products, the value of h computed from Equation 21 is 650 cal/g total volatile matter. Although this example was obviously contrived, a reasonable variation in the assumed distribution that is consistent with experimentally observed devolatilization yields for high-volatility coals would not result in a greatly different value for h_v.

TABLE 3. DISTRIBUTION OF VOLATILES FOR A RAW BITUMINOUS COAL

Coal Component Devolatized	Total g/g (C + H + O)	Volatile Species	Total g/g (C + H + O)	Heat of Hydrogenation to CH_4 (cal/g Volatile Species)
Carbon	0.67	CO	0.15	121
Hydrogen	0.14	H_2O	0.12	
Oxygen	0.19	CH_4	0.17	
		C_2H_6	0.13	592
		C_6H_6	0.30	1850
		H_2	0.04	
		C(tar)	0.10	
			1.01	

The heat of reaction of the fixed carbon in coal chars with reactive gases has been treated in different ways by different investigators. For example, the assumed value of h_f in Figure 4 is consistent with the assumption that the standard heat of formation of carbon in coal chars is the same as that of graphite. This simplifying assumption is most commonly employed in calculating the heats of reaction in coal-char-gasification systems. It has also been proposed, however, that the carbon in coal chars is amorphous, and that by analogy with other types of amorphous carbon, coal-char carbon should have a higher heat of formation than graphite.

Stephen [35], in computing thermodynamic characteristics of subbituminous coal-char-gasification systems, assumed that coal-char carbon has a standard heat of formation that is 2.6 kcal/g-atom carbon (216 cal/g carbon) larger than that of graphite. This value is consistent with the extreme values for standard heats of formation of amorphous carbon at 298°K that have been reported in the literature [36] and having values of standard heats of formation of carbon that can be deduced from some of the generalized combustion-heat formulas expressed as a function of coal composition.

Accurate determination of the standard heat of formation of a raw coal is particularly difficult. Although such a value can be theoretically computed from experimental determinations of the heats of carbonization, such a procedure is complicated by the need to know or estimate carbonization product yields and the thermodynamic properties of these products. As indicated previously, this information is neither complete nor well defined, even for a few coals. Therefore, the more usual procedure in computing heat effects for gasification reactions in which coal is converted to coal char is to make use of the heats of combustion, as determined experimentally or taken from one of the many published general correlations [37, 38]. The main disadvantage of this procedure, however, is the error that can result from computing a relatively small number as the difference of two much larger numbers that are of limited accuracy. In addition, very little information is available for estimating heats of combustion of partially devolatilized coal residues as a function of composition and preparation conditions—information that would be of value in detailed analyses of coal-gasification systems.

PHYSICS OF GASIFICATION SYSTEMS

Both physical and chemical processes are important in coal and coal-char gasification. In a limited sense, physical processes can be defined as those that affect the practical operability of an integral conversion system as well as the conditions under which chemical reactions occur. Such processes include those that affect mass transfer, heat transfer, pressure drop, other coal hydrodynamic behavior, physical properties of solids undergoing gasification, and solids agglomeration and attrition. Much of the information necessary to quantitatively characterize certain of these processes as they pertain to coal gasification systems as well as to other types of systems is available in a voluminous literature, including textbooks and handbooks. Although most of this information is not amenable to review within the scope of this book, some areas that are particularly important to modern coal-gasification concepts have received considerable attention in recent years and are discussed here. Two of these areas include physical processes related to fluidized-bed systems and the characterization of coal and coal-char structural variations during gasification. Characterization of the physical processes that occur in fluidized-bed contacting systems is important to the design of most modern gasification systems currently being developed; moreover, coal and coal-char structural characteristics are important parameters that affect the behavior of any contacting system, including fluidized beds. Although fixed beds and suspension systems are also of interest in coal gasification, the physical characterization of such systems has been described in various reviews [8, 39, 40].

Characteristics of Gas-fluidized Beds

Aggregative gas-fluidized beds can be generated by passing a gas upward through a bed of relatively fine particles. At low flow rates the bed remains stationary (i.e., fixed bed) as the gas passes through voids between the particles. As the gas flow rate increases, the pressure drop across individual particles and across the total bed of solids also increases. Ideally, a specific flow rate is eventually reached at which the particles are just suspended in the upward-flowing gas, and the pressure drop through any section of the bed is approximately equal to the weight of particles in that section. At this point the net vertical components of compressive force between adjacent particles disappear, and the bed is considered to be incipiently fluidized, or at a minimum fluidization. The superficial gas velocity at this condition, referred to the empty-tube plug-flow value, is defined as the minimum fluidization velocity, V_{mf}. The average interstitial gas velocity in the fluidized bed, V_f, is

$$V_f = \frac{V_{mf}}{\epsilon_{mf}} \qquad [22]$$

where ϵ_{mf} is the average void fraction in the bed at minimum fluidization.

As the gas velocity is increased beyond minimum fluidization, the total pressure drop remains relatively constant, but gas bubbles tend to form and pass upward through the bed, as depicted in Figure 8. There are two major regions of solids density in a bubbling fluidized bed, usually referred to as the emulsion phase (high solids density) and the bubble phase (low solids density). It is often

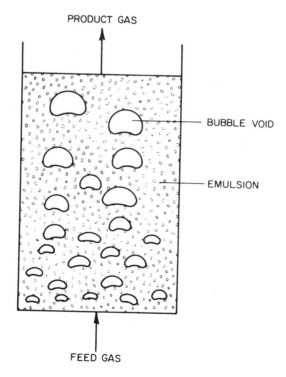

Figure 8 Schematic diagram of bubbling fluidized bed.

22 FUNDAMENTALS OF COAL GASIFICATION

assumed, as a first approximation, that all gas in excess of that needed to bring the bed to incipient fluidization passes through the bed as bubbles, whereas the emulsion phase remains at minimum fluidizing conditions.

Bubbles formed at the bottom of a bed do not remain the same size as they pass upward through the bed, although this is often assumed in analyses of fluidized-bed behavior. Instead, bubbles can grow by coalescing with other bubbles or by splitting under conditions of instability. The rate of bubble growth generally increases with increasing bubble diameter. A particularly important effect of bubble interactions in the bed is the agitation and relatively large solids mixing rates that result. This feature is particularly desirable in coal-gasification systems in which control is simplified by the approximately isothermal conditions that result from large solids circulation rates, even with the highly exothermic and endothermic reactions that can occur.

Other advantages of the fluidized bed for coal gasification include relatively high gas-solids heat-transfer rates and the ease of controlling the processes due to the fluid-like flow of particles, which facilitates large-scale operation. Potential disadvantages of the fluidized bed include gas bypassing of the solids, nonuniform solids residence times for continuous solids feed and withdrawal, gaseous backmixing, solids attrition and agglomeration, and solids elutriation. Solids elutriation becomes a problem particularly when a relatively large distribution of particle sizes is used, which causes fine particles to be carried out of the bed with the product gas.

A major objective is elucidating the behavior of fluid beds is to understand the movement of gases and solids. A significant breakthrough in clarifying these processes was made by Davidson and Harrison [41] in 1963. Their model was later extended by Murray [42], with an alternative treatment given by Jackson [43]. Based on these models, it is now understood that when the bubble velocity is greater than the interstitial gas velocity in the emulsion phase, there is a convective circulation of gas from the top of the bubble void into the emulsion phase. The gas is then swept around and returned to the bubble void near its lower extremity. The region around the bubble through which this circulation occurs is called the *bubble cloud*, as schematically depicted in Figure 9. Significantly, a turbulent mixing zone just below the bubble void causes particles of solids to be drawn up behind the rising void and to form a wake, also shown in Figure 9. Solids in this wake move up with the bubble void but are also continuously exchanged with emulsion solids, thus resulting in a primary mechanism for solids mixing in a vigorously bubbling bed. Convective movement of the solids also occurs by displacement of the emulsion phase by rising bubbles. Solids in the wake carried to the surface of the bed fall back and lie on top of the bed, leading to a net downward mass flow of solids in the emulsion phase of a batch-solids system to compensate for the upward mass flow of wake solids. In systems where solids are continuously fed and withdrawn from the bed, the

PHYSICS OF GASIFICATION SYSTEMS 23

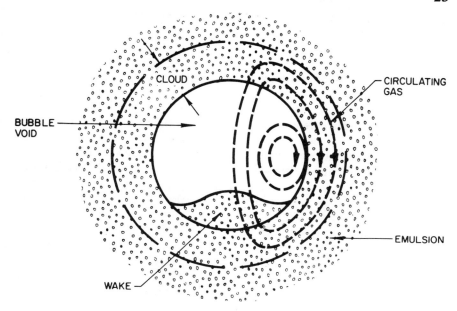

Figure 9 Simplified diagram of bubble region.

direction and mass flow rate of emulsion solids depend on a number of factors, including solids feed and withdrawal locations.

As the bubble velocity:interstitial emulsion gas velocity ratio decreases, the cloud thickness increases until the bubble velocity is lower than the interstitial emulsion gas velocity. At this point, in addition to a circulating pattern of gas flow between bubble void and emulsion, there is also a flow of emulsion gas into the bottom and out of the top of the bubble void. This "through-flow" gas does not directly circulate back to the bubble void. Diffusive-type gas transfer can also occur between bubble void and cloud and between cloud and emulsion, as another mechanism for net transfer of gas from bubble void to emulsion.

Despite the significant advances made in the last decade or two in elucidating some of the complex processes involves in fluidized-bed technology, there are still important gaps in information and theory that prevent a priori prediction of fluidized-bed behavior. Recent reviews have indicated the need for more information on bubble inception, the effects of gas-distributor design, the phenomenon of bubble coalescence, other bubble interactions, and bubble distribution in large beds [44-48]. Idealized models of varying sophistication have, however, been proposed to describe behavior in fluidized beds for a limited

24 FUNDAMENTALS OF COAL GASIFICATION

range of conditions. A consideration of the different types of models proposed is, therefore, of interest in evaluating available techniques for designing coal-gasification fluidized-bed reactors.

Fluidized-bed Models. The earliest models proposed to describe the gas-solid contacting configurations in fluidized beds did not reflect the existence of bubbles. Such models idealized the system by assuming isothermal conditions and by various combinations of assumptions with respect to perfect backmixed or plug flow of solids and perfect backmixed or plug flow of gas. Although such models (which are still extensively used) have been found to adequately describe experimental results for limited conditions, a variety of models has also been proposed that either directly or indirectly recognize the multiphase nature of bubbling fluidized beds and account for a number of mass and heat transfer mechanisms.

These multiphase models can be broadly classified into two major categories: simple two-phase models and bubbling-bed models. For simple two-phase models (Figure 10), the fluidized bed is considered to be composed of two parallel, single-phase regions, one of dense solids concentration and one with dilute or zero solids concentration. Individual gas streams pass through each of

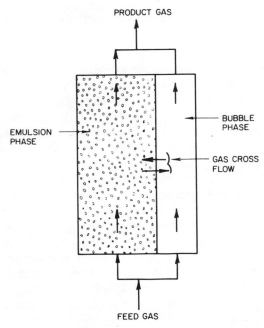

Figure 10 Schematic diagram of simple two-phase models.

these phases, and there is also a cross-flow of gases between phases. With this type of model the dense-solids-concentration region is associated with the emulsion phase and the lean-solids-concentration region, with the bubble phase.

A summary of the assumptions made in different studies to characterize simple two-phase models is shown in Table 4, which was adapted from very thorough surveys made by Grace [44] and by Rowe [46]. In most of the models considered in Table 4, parallel gas flow occurs in the emulsion and bubble phases, although in some models net gas flow in the emulsion phase is assumed to be zero. In the bubble phase, plug flow is usually assumed, whereas assumptions concerning gas-flow patterns in the emulsion phase have varied from plug flow to perfect mixing, with finite backmixing represented by both axial dispersion and tanks-in-series models. Although some of the simple two-phase models account for the presence of solids in the bubble phase, relatively little consideration has been given to solids-mixing patterns or residence-time distributions, since most of the interest has been in gas–solids catalytic reactions and not in heterogeneous gas–solids reactions, which are important in coal and coal-char gasification.

Simple two-phase models provide a useful engineering approach for designing or scaling up fluidized-bed reactors, based on experimental evaluation of the parameters inherent in a given model. Because of the level of empiricism involved in using this type of model, however, extrapolation of laboratory data to design large-scale reactors in commercial goal gasification systems necessarily involves a high degree of uncertainty in projecting performance. Thus for coal-gasification systems, there may be an advantage to using the more complex bubbling-bed models for scale-up, since these models, based at least in part on characterization of fundamental fluid-mechanical processes, make it possible to account for phenomena that may occur on a large scale but that cannot be easily anticipated from empirical descriptions of laboratory data.

Grace [44] and Rowe [46] have also made thorough reviews of the more complex bubbling-bed models depicted in Figures 8 and 9; a summary of the assumptions made in the various proposed models is shown in Table 5, adapted from these authors. The models are all based on the properties of single rising bubbles, often using results from independent theoretical or experimental studies to account for bubble rise velocity, cloud formation, through flow of gas in bubbles, wake characteristics, and other phenomena. For most of the models listed in Table 5, unknown parameters can be selected to fit experimental reaction data. Although some of the parameters can be at least approximately determined on the basis of fundamental considerations, as indicated previously, there is not enough information to accurately predict bubble size distributions and bubble interaction effects. For this reason an adjustable parameter that is often selected to fit experimental results is the effective bubble diameter.

TABLE 4 CHARACTERISTICS OF SIMPLE TWO-PHASE MODELS

Reference	Bubble Gas Flow[a]			Solids in Bubble Phase			Gas Flow in Emulsion Phase				
	$Q_e = 0$	$Q_b = Q_0 - Q_{mf}$	Not Stated	Yes (Dilute)	No	Plug Flow	Perfect Mixing	Axial Dispersion	Tanks in Series	Down flow
49	–	x	–	–	x	x	x	–	–	–
50	–	x	–	x	–	x	–	–	–	–
51	x	–	–	x	–	x	x	–	–	–
52	–	x	–	–	x	–	–	x	–	–
53	–	–	x	x	–	x	–	–	–	x
54	–	x	–	–	x	x	–	–	–	–
55	x	–	–	–	x	–	–	x	–	–
56	–	x	–	–	x	–	–	–	x	–
57	–	x	–	x	–	x	x	–	–	–
58	–	x	–	–	x	–	–	x	–	–
59	–	–	x	–	x	x	–	–	–	x

[a]Values: Q_{mf} = gas flow rate at minimum fluidization; Q_0 = total gas flow rate; Q_b = gas flow rate in bubble phase; Q_e = gas flow rate in emulsion phase.

TABLE 5 CHARACTERISTICS OF BUBBLING-BED MODELS

Reference	Bubble Gas Flow[a]			Bubble Phases			Gas Flow in Emulsion Phase				Gas Exchange Resistance[b]		Model		Bubble Detail			
	$Q_e = 0$	$Q_b = Q_0 - Q_{mf}$	$Q_b = Q_p - Q_{mf}Q_t$	Simple Empty	Empty + Cloud (One Phase)	Empty + Cloud (Two Phase)	Plug Flow	Perfect Mixing	Axial Dispersion	Tanks in Series	Bubble-Cloud Interface	Cloud-emulsion Interface	Davidson	Murray	Through Flow	Mixing	Wake	Growth
60	x	x	—	x	—	—	x	x	—	—	Int.	o	—	—	—	x	—	—
41	x	x	—	x	—	—	x	x	—	—	Int.	o	—	—	—	x	—	—
61	x	x	—	—	x	—	x	—	—	—	o	o, ∞	x	—	—	x	x	—
62	x	x	—	—	x	—	x	—	—	—	o, Int.	∞	x	—	—	x	x	—
63	x	x	—	—	x	—	x	—	—	—	o	Int.	—	x	—	x	x	x
64	x	x	—	x	x	—	x	x	x	—	Int.	o, int, ∞	x	—	—	x	x	—
65	x	x	—	x	—	x	—	—	—	—	Int.	Int.	—	—	—	x	—	—
66	—	—	x	—	—	—	—	—	—	—	Int.	Int.	—	x	x	x	x	—
67	—	x	—	—	x	—	—	—	x	—	Correlating	—	x	—	x	x	—	x
68	—	x	—	—	—	—	x	—	—	x	Correlating	—	x	x	—	x	x	—

[a]Values: Q_{mf} = gas flow rate at minimum fluidization; Q_0 = total gas flow rate; Q_b = gas flow rate in bubble phase; Q_s = gas flow rate in emulsion phase; Q_t = "through-flow" gas rate in bubble.
[b]Abbreviations: o = no mass-transfer resistance; int = intermediate mass-transfer resistance; ∞ = infinite mass-transfer resistance (no transfer); correlation = net transfer rate between bubble and emulsion kept as correlating parameter.

28 FUNDAMENTALS OF COAL GASIFICATION

In most experimental studies of fluidized beds, models have been evaluated in catalytic reactor systems by measuring the concentrations of the reactants and products in the inlet and outlet gas streams. However, Chavarie and Grace [47] have recently compared the predictions of some of the bubbling-bed models not only with total reaction yields, but also with internal gas concentration gradients within the bubble and emulsion phases. In the present study a detailed experimental investigation was made of the catalytic decomposition of ozone in a two-dimensional fluidized bed using porous alumina particles. The fluidized-bed models considered were those of Orcutt, Davidson, et al. [60], Partridge and Rowe [63], Kunii and Levenspiel [66], and Kato and Wen [67]. Of the four, Kunii and Levenspiel's model was found to best fit the internal gas-concentration profiles and net gas yield for the range of experimental conditions investigated. This was true even when modifications were made in the model-derived values of parameters to improve agreement. It is interesting to note that measured mean bubble sizes halfway up the fluidized bed used by Chavarie and Grace were surprisingly close to calculated effective bubble sizes necessary to optimize the Kunii-Levenspiel model. However, the analysis made was for a two-dimensional bed, and radial variations in properties were not considered in comparative evaluations. Thus it is uncertain to what extent this analysis is meaningful in terms of large-scale three-dimensional fluidized beds.

General Properties of Fluidized Beds. Quantitative empirical correlations and data tabulations useful in the design of fluidized-bed systems have been obtained in various investigations. Such correlations are valuable both for estimating gross bed behavior and for providing a basis for estimating more detailed properties necessary to apply certain of the fluidized-bed models discussed previously.

Particle-shape factors. Shape factors for different nonspherical coal particles are often needed for the application of empirical correlations to estimate properties in fluidized beds. The shape factor, ϕ, is defined as the ratio of the surface area of a sphere to the surface area of a particle, both having the same volume. Thus the shape factor ϕ for a sphere is equal to unity, and for any other shaped particle, $0 < \phi < 1$. Table 6 lists shape factors determined for a variety of coals ranging in rank from anthracite to subbituminous. The results of Austin, Gardner, et al. [69] are of particular interest in that shape factors were found to be relatively independent of particle size for particle diameters ranging about 40–1200 μ. These investigators also found that the shape factor varied systematically with coal volatile matter, with maximum sphericity at a volatile matter content of about 17.9% maf.

Bed voidage at minimum fluidization. The bed voidage fraction at minimum fluidization, ϵ_{mf}, is also an important parameter in designing fluidized beds, particularly since it is often assumed that the properties of the emulsion phase,

TABLE 6 SHAPE FACTORS OF DIFFERENT COALS

Material	Volatile Matter (maf%)	Shape Factors	Reference
St. Nicholas (Pa.)	4.5	0.64	69
Lorree Raven (Pa.)	5.4	0.61	69
Dorrance (Pa.)	5.8	0.59	69
Treverton (Pa.)	9.0	0.69	69
Upper Freeport (Md.)	17.9	0.83	69
Upper Kittanning (Pa.)	18.0	0.78	69
Pratt (Ala.)	29.2	0.75	69
Lower Kittanning (Pa.)	30.2	0.71	69
Upper Freeport (Pa.)	33.0	0.70	69
Cobb (Ala.)	33.1	0.68	69
Pittsburg (Pa.)	37.4	0.65	69
Glen Mary (Tex.)	39.4	0.68	69
Pittsburgh (Pa.)	39.5	0.65	69
Pittsburgh (Pa.)	40.0	0.62	69
Pittsburgh, No. 8 (Ohio)	42.4	0.62	69
Illinois No. 6 (Ill.)	45.4	0.62	69
Anthracite	–	0.63	70
Bituminous	–	0.63	66
Coal dust (natural up to 3/8 in.)	–	0.65	71
Coal dust (pulverized)	–	0.73	71

even in bubbling beds, are the same as those at minimum fluidization. The value of ϵ_{mf} has been found both to decrease [72, 73] and to increase [74–77] with increasing particle size for coal-related materials. Correlations of this parameter with other potentially significant factors such as shape factors, particle densities, and particle size distributions are not available. Kunii and Levenspiel [66] have suggested that since ϵ_{mf} is only slightly larger than voidage in a packed bed, values can be estimated from random packing data or can be measured experimentally.

Minimum-fluidization velocity. A variety of correlations proposed to predict minimum-fluidization velocities has been reviewed by Leva [78], Wen and Yu [79], and Frantz [80]. Many of the correlations proposed are based on the definition that minimum fluidization occurs when the total drag force across the bed (pressure drop times cross-sectional area of tube) is equal to the weight of solids in the bed. Wen and Yu [79] and Kunii and Levenspiel [66] have adopted this procedure, using the fixed-bed pressure drop correlation of Ergun [81] to develop the expression

$$\frac{1.75}{\varphi \epsilon_{mf}^3} \left(\frac{D_p V_{mf} \rho_g}{\mu} \right)^2 + \frac{150(1 - \epsilon_{mf})}{\varphi^2 \epsilon_{mf}^3} \left(\frac{D_p V_{mf} \rho_g}{\mu} \right) = \frac{D_p^3 \rho_g (\rho_s - \rho_g) g}{\mu^2}$$

[23]

where

D_p = average particle diameter (in cm)
ρ_g = gas density (in g/cm^3)
ρ_s = solids particle density in (g/cm^3)
μ = gas viscosity in (g/cm-sec)
g = gravitational constant (in cm/sec^2)

Although the fixed-bed pressure drop equation developed by Ergun was for uniformly sized solids, it, as well as the related minimum-fluidization equation given above, is often applied to systems with a particle size distribution based on the approximation

$$\frac{1}{D_p} = \sum_i \frac{X_i}{d_{p_i}} \qquad [24]$$

where

X_i = weight fraction
d_{p_i} = average particle size of each weight fraction (in cm)

For cases where ϵ_{mf} and ϕ are unknown, Wen and Yu [79] recommend the following modification of Equation 23. They found that for a wide variety of systems:

$$\frac{1}{\varphi \epsilon_{mf}^3} \cong 14; \quad \frac{1 - \epsilon_{mf}}{\varphi^3 \epsilon_{mf}^3} \cong 11 \qquad [25]$$

and that substitution of these expressions in Equation 23 gives

$$V_{mf} = \frac{\mu}{D_p \rho_g} \left\{ \left[(33.7)^2 + 0.0408 \frac{D_p^3 \rho_g (\rho_s - \rho_g) g}{\mu^2} \right]^{1/2} - 33.7 \right\} \qquad [26]$$

Although Wen and Yu found Equation 26 to be in reasonable agreement with a wide variety of experimental data, particularly at relatively low pressures, Knowlton [81] found a significant disagreement between predictions of this expression and results obtained on fluidization of coals and related materials at pressures up to 70 atm. In a study conducted at the IGT to prepare a technical coal data book [31], it was proposed to use a set of values for the parameters $(1/\varphi \epsilon_{mf}^3)$ and $(1 - \epsilon_{mf})/\phi^2 \epsilon_{mf}^3$ different from those recommended by Wen and Yu. This proposal came after analysis of fluidization data obtained with coal-related materials. The IGT values

$$\frac{1}{\varphi \epsilon_{mf}^3} = 8.81, \quad \frac{1 - \epsilon_{mf}}{\varphi^2 \epsilon_{mf}^3} = 5.19 \qquad [27]$$

when substituted into Equation 23 lead to

$$V_{mf} = \frac{\mu}{D_p \rho_g} \left\{ \left[(25.25)^2 + 0.0651 \frac{D_p^3 \rho_g (\rho_s - \rho_g) g}{\mu^2} \right]^{1/2} - 25.25 \right\}$$

[28]

Equation 28 was found to agree with minimum-fluidization velocities obtained in the studies [82, 83] described in Table 7 with a standard deviation of 30%.

The ratio of minimum-fluidization velocities predicted by Equation 28 to velocities predicted by Equation 26 ranges from about 2 at low Reynold's numbers to about 1.25 at high Reynold's numbers. These variations may be due in part to differences in the range of conditions and types of materials considered in development of the two correlations.

Bed expansion. When superficial velocities of the gas entering a fluidized bed are increased beyond the point required for minimum fluidization, the bubbles that form and pass through the bed lead to bed expansion. Therefore, it is of interest to determine the extent of this expansion, as a function of operating conditions, to facilitate design of fluidized beds. Experimental determination of bed expansion, however, is not a straightforward procedure. This is because gas bubbles that erupt from the surface of a fluidized bed tend to throw solid particles up above the bed, some of which are carried out of the reactor (elutriation) and some of which fall back into the bed. As a result, there is not necessarily a sharp demarcation between the top of a fluidized bed and the lean gas phase above the bed, which can lead to differences in the interpretation of bed-expansion measurements.

Numerous correlations for bed expansion in fluidized beds have been proposed, and some studies have found that bed expansion tends to decrease with increasing tube diameter [39, 84, 85]. One empirical correlation proposed in the IGT study for the coal-data book [32] and based on results obtained primarily with coal-related solids, is as follows. For $D_T \leqslant 6.35$ cm:

$$\frac{L_f}{L_{mf}} = 1 + \frac{0.437 (V - V_{mf})^{0.57} \rho_g^{0.083}}{\rho_s^{0.166} V_{mf}^{0.063} D_T^{0.445}} \qquad [29]$$

TABLE 7 SOURCES OF DATA FOR MINIMUM-FLUIDIZATION VELOCITY CORRELATION[a]

Reference	Bed Diameter (cm)	Fluidized Solids	Particle Diameter (cm)	Particle Density (g/cm^3)	Fluidizing Medium	Operating Pressure (atm)
72	4.2	Coal	0.031–0.056	1.31–1.33	Air	1
74	8.5	Coke	0.009–0.066	1.80	H_2, air CO_2, argon	1
76	2.5, 5.1	Lignite–char dolomite, periclase	0.007–0.044	0.82–3.56	N_2, H_2, CO_2	1
77	9.4	Char	0.013–0.287	0.26–0.37	CO_2	1
83	6.4	Siderite	0.012–0.021	3.93	Air, CO_2, H_2, freon	1–69
82	29.2	Coal, char lignite, siderite	0.024–0.029	1.17–3.93	N_2	1–69

[a] All data obtained at 21°–27°C.

TABLE 8 SOURCES OF DATA FOR BED-EXPANSION CORRELATION[a]

Reference	Bed Diameter (cm)	Fluidized Solids	Particle Diameter (cm)	Particle Density (g/cm)	Fluidizing Medium	Operating Pressure (atm)	Range of V/V_{mf}	Range of L_f/L_{mf}
76	2.5, 5.1	Lignite, char, dolomite, periclase	0.007–0.044	0.82–3.56	N_2, H_2, CO_2	1	1–20.5	1–2
77	5.1	Char	0.013	0.37	CO_2	1	1–12.0	1–1.53
83	6.4, 14.0, 29.2	Siderite	0.007–0.036	3.93	Air, freon, N_2	1–69	1–10.5	1–1.49
82	29.2	Lignite, coal, char, siderite	0.024–0.029	1.17–3.93	N_2	1–69	1–2.6	1–1.12
86	25.4	Coal	0.005–0.020	1.39	Air	1	1.3–40.2	1–1.63
87	30.5	Char, limestone	0.046–0.216	1.36–2.64	Air	1	1.2–7.3	1–2.15

[a] All data obtained at 21–27°C.

For $D_T > 6.35$ cm:

$$\frac{L_f}{L_{mf}} = 1 + \frac{1.95(V - V_{mf})^{0.738} D_p^{1.006} \rho_s^{0.376}}{V_{mf}^{0.957} \rho_g^{0.125}} \qquad [30]$$

where

L_{mf} = bed height at minimum fluidization (in cm)
L_f = bed height (in cm)
ρ_s = average particle density of solids (in g/cm^3)
ρ_g = average gas density (in g/cm^3)
D_T = tube diameter (in cm)
V = superficial feed-gas velocity referred to empty tube plug flow value (in cm/sec)

Note that for $D_T \leq 6.35$ cm, the expansion ratio, L_f/L_{mf}, decreases with increasing tube diameter, whereas for $D_T > 6.35$ cm, the expansion ratio is independent of the tube diameter. Equations 29 and 30 were fitted to 90% of the data obtained in the studies [86, 87] described in Table 8, including high-pressure data, with a maximum deviation of 12%.

Solids circulation. In a fluidized bed, gross solids are circulated upward primarily by being transported in the wake of rising bubbles, and in a downward direction by compensating flow in the emulsion phase. Talmor and Benenati [88] have studied the movement of solids in bubbling-air fluidized beds by a tracer technique, which was cation-exchange resin in hydrogen and sodium forms. Beginning with a layered bed of tracer and nontracer, they fluidized the bed with air for varying lengths of time and determined the amount of tracer solid that crossed the original interface. The solids circulation flux, J, was determined from the experimental results based on a differential equation given by Katz and Zenz [89] and was correlated with the following expression for the range of conditions employed:

$$J = 0.785 (V - V_{mf}) \exp(-66.3 D_p) \qquad [31]$$

where

J = solids circulation flux referred to tube cross-sectional area (in g/cm^2-sec)
D_p = average particle diameter (in cm)

Talmor and Benenati [88] showed that Equation 31 correlates the data of Koble [90] and of Leva and Grummer [91]. Woolard and Potter [92] subsequently demonstrated that Equation 31 also fits the data of Bart [93] and of Vakhrushev and Erokin [94], with the result that the correlation is applicable on the following range of parameters:

Bed diameter	3.2–10.2 cm
Particle diameter	0.0067–0.065 cm
Particle density	0.75–2.65 g/cm^3
Size distribution	mixed and uniform
Solids	microspheres, round sand, sharp sand, silica gel ion-exchange resins, cracking catalyst, coke

Thus, as an empirical representation, Equation 31 appears to describe circulation data obtained over a relatively wide range of conditions, at least at atmospheric pressure. Kunii and Levenspiel [66] have suggested, however, that in view of the complicated interactions in fluidized beds, the relatively simplistic basis used for deducing solids fluxes from experimental tracer studies should limit the application of Equation 31 to first-order approximations.

In addition to the net upward and downward solids fluxes in fluidized beds, there is also a cross-flow between wake and emulsion solids. Kunii and Levenspiel, in developing their fluidized-bed model, recommended the correlation of Yoshida and Kunii [95] to describe this cross-flow; in other words:

$$K^* = \frac{\text{(volume of solids transferred between the wake and emulsion solids)}}{\text{(volume of bubble)(time)}}$$

$$\approx \frac{3(1 - \epsilon_{mf})V_{mf}}{(1 - \delta)\epsilon_{mf}D_b} \qquad [32]$$

where

δ = volume fraction of bubbles in bed
D_b = characteristic bubble diameter

Pore Structural Characteristics of Coals and Coal Chars

Variations in the pore structure of coal and coal-char particles that occur during gasification are important in considering hydrodynamic aspects of integral gas-solid contacting systems as well as in evaluating gaseous mass-transfer behavior and available reactive surface areas within individual particles. Coals and coal chars exhibit complex and unique pore structures, with a wide distribution of pore sizes. Systematic analysis of coals and coal-derived solids is complicated by the heterogeneous nature of such substances with respect both to organic and mineral constituents. Thus the total pore size distribution for a "whole" coal represents the sum of the distribution characteristics of relatively homogeneous organic constituents (macerals), the distribution characteristics of mineral constituents, and the pore volume contributions of interface regions between the various phases present.

Anderson, Hall, et al. [96], in a pioneering study of the pore structures of a number of American coals, found total porosities ranging 2.5-18%, and observed that a significant fraction of pores had openings (diameters) of 4.9-5.6 Å. In a later study, Gan, Nandi, et al. [97, 98] characterized the pore structure of a wide range of American coals in terms of porosities divided into three pore size ranges: (1) micropores, which are accessible through effective pore diameters of less than 12 Å, (2) transitional pores, with diameters of 12-300 Å, and (3) macropores, with diameters greater than 300 Å. It was also found in this study that the micropores can represent a significant fraction of pore volume.

Walker and Hippo [99] have discussed some of the fundamental aspects relevant to characterizing the pore structure of chars and their precursor coals, based largely on the detailed X-ray study of Cartz and Hirsch [100]. They pointed out that coals are composed of aromatic and hydroaromatic layers, terminated at their edges by various functional groups and cross-linked by different functional groups. The average size of the layers and the number of layers that are aligned tend to increase with increasing coal rank. Poorer alignment between packets of layers produces internal porosity, which makes coal a microporous material. On heating, coals lose volatile matter, primarily from the periphery of the layers. Some coals pass through a softening or plastic stage before solidifying into a coke or char. The extent of softening, which does not occur to a significant extent with low- or high-rank coals, is thought to be largely dependent on the concentration and thermal stability of cross-linking groups. The micropore structure of the precursor coal is preserved in chars that are not heated to a very high temperature; in fact, microporosity can become even more accessible in chars because of the loss of volatile matter. At very high temperatures, however, microporosity is rapidly lost due to thermal breakage of the cross-links between planar regions in the char, allowing improved alignment of these regions.

With respect to their structure, Walker and Hippo [99] envision chars as being composed of small trigonally bonded planar regions, terminated by much fewer heteroatomic functional groups than present in the original coal and still containing substantial cross-linking, although less than in the original coal. They also suggest that the exact structure of a char produced from a given coal will depend on a number of variables such as coal pretreatment, coal particle size, heating rate, maximum temperature, time at maximum temperature, gas composition, and total pressure during heating. Based on differences in the distribution of pore volume with respect to pore size (diameter), chars from coals of different rank would be expected to exhibit differing degree of inhibition to internal mass-transfer effects during gasification. This is related to the fact that the internal surface area of coals and chars exists almost solely in the micropores, where gaseous diffusion is slow. Diffusion is, in fact, an activated process for pore diameters less than about 5 Å, as shown by Walker, Austin, et al. [101]. Walker and Hippo [99] proposed that to efficiently use the active sites located in the micropores for reaction, a large number of feeder pores (macro and transitional) to which the micropores are connected is required to limit diffusion length within micropores. Diffusion through feeder pores to the mouth of micropores would be reasonably rapid; therefore, gaseous reactant concentrations at the mouths of micropores could closely approach reactant concentrations at the exterior particle surface. Furthermore, since the results of Gan, Nandi, et al. [97, 98] have shown that lower-rank coals tend to have a greater percentage of the pore volume in larger pores than do higher rank coals, it can be expected that chars derived from lower rank coals would have a larger feeder pore system and be less limited with respect to mass transport during their gasification.

Because of the complex pore structure of coals and coal chars, many experimental techniques have been employed to characterize their physical properties in terms of solids densities, total pore volumes, pore volume distributions, and internal surface areas. Useful information has been obtained by density measurements and sorption studies using different fluids at various temperatures.

Solids Densities of Coals and Coal Chars. Helium displacement is used most often to determine solids densities or true densities since the helium is not easily adsorbed on the internal surfaces of coal at room temperature, and it is believed to penetrate pore opening diameters greater than 3-4 Å [102, 103]. With some highly adsorptive carbons, some adsorption of helium has been found at low temperatures [104, 105]. For this type of material, it is recommended that helium density measurements be made at elevated temperatures of about 300°C [106].

Coal and coal-char true densities measured by helium displacement and expressed on an maf basis have shown a strong correlation strongly with the hydrogen content of the material [31, 97, 107]. In the IGT study for the coal-data book [31] the following correlation was developed to describe the relationship between the densities reported in the literature [96, 97, 102, 107-115] for a variety of coals; carbonized coals; high-temperature cokes; and maceral concentrates, including exinite, fusinite, and vitrinite:

$$V_{He} = \frac{1}{\rho_{He}} = 0.4397 + 0.1223 \, Y_H - 0.01715 \, Y_H^2 + 0.001077 \, Y_H^3 \quad [33]$$

where

Y_H = hydrogen content of coal or char (maf) (in mass%)
V_{He} = specific volume in helium [in cm³/g (maf)]
ρ_{He} = "true" density in helium [in g/cm³ (maf)]

This correlation, applied to data with Y_H ranging about 0–7.7, represents specific volumes or densities of the organic solids on an maf basis. In view of the arbitrary form of the equation, extrapolation should be avoided. To obtain the specific volume of the solids, including ash, the following relationship can be applied:

$$V'_{He} = V_{He}(1 - X_A) + \frac{X_A}{\rho_A} \qquad [34]$$

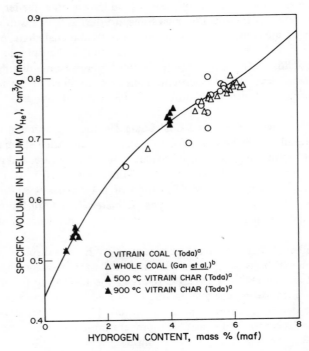

Figure 11 *Correlation of true specific volume of coals and coal-derived solids as a function of hydrogen content.*

where

X_A = ash mass fraction
ρ_A = true density of ash in g/cm³
V'_{He} = specific volume of coal or coal-char solids in cm³/g

An average value for ash density of about 2.7 g/cm³ is often used to represent coal density data on an ash-free basis [97, 116]. Of course, experimental values of ash or mineral densities are preferred, particularly if there is a large concentration of high-density substances such as pyrite or marcasite [31]. Equation 33 was developed on the basis of a set of 190 data points, with a standard deviation of 0.012 cm³/g over the whole range. Figure 11 [117] compares the correlation with some of the data considered in its development.

Pore Volumes of Coals and Coal Chars. Density measurements obtained by mercury displacement have been used in conjunction with helium density measurements to obtain information on the total pore volumes and pore volume distributions of coals and chars. When a porous solid material is immersed in mercury at a given pressure P, the mercury will penetrate all pores with an entry diameter greater than some minimum value, D_m, depending on the pressure. For circular pores, the relationship between minimum pore diameter and pressure is usually computed from the Washburn equation [118]:

$$D_m = - \frac{\sigma \cos \theta}{P} \quad [35]$$

where

D_m = minimum pore diameter (for pores having circular entries)
P = pressure
σ = surface tension of mercury
θ = contact angle between mercury and the material forming the opening

Using reasonable values of σ and θ at room temperature, Equation 35 reduces to

$$D_m = \frac{120.8}{P} \quad [36]$$

for D_m in angstroms and P in atmospheres. Using Equation 36, measurements made over a pressure range of 1 atm (often used to estimate total porosities) to

40 FUNDAMENTALS OF COAL GASIFICATION

as high as 1000 atm (used in some studies of macropore distributions) will result in a range of D_m values of 120.8-121 Å. The specific volume measurements, V_{Hg}, made by mercury displacement are equal to the sum of the specific volume of solids plus the volume of pores accessible through openings with a diameter less than D_m. Therefore, assuming that helium displacement measurements give the true specific volume of the solids, the void volume accessible through pore openings having diameters less than D_m can be determined from the expression

$$V_{D_m} = V_{Hg} - V_{He} \qquad [37]$$

The total pore volume between two values for the pore-opening diameter can be determined without helium density measurements by measuring the difference in mercury penetration into the solid between the two pressure levels corresponding to the two pore-opening diameters. Some caution is necessary, however, in interpreting mercury displacement results, in terms of pore volumes, at low and high hydrostatic pressures. At low pressures, some mercury penetration can be due to the filling of voids between particles. The pressure at which interparticle filling can occur increases with decreasing particle size [97]. It has been recommended that the limiting pressures for interparticle filling be determined experimentally with nonporous materials of the same particle size as the porous solids [97]. Interparticle filling may also be accounted for by calibration. In addition, at very high pressures the compressibility of the coal or coal-char material has to be considered, or else apparent pore volumes will be too high [37].

Total pore volumes of a variety of American coals and of Japanese and American vitrains are shown in Figure 12. The results are represented on an ash-free basis and assume that all of the pore volume is associated with the organic material and not with the ash. The American coals used by Gan, Nandi, et al. [97] (315 μ average particle diameter) varied in vitrinite content from about 62 to 96 vol%, and in ash content from about 2 to 25 mass%. The mercury densities used by Gan and colleagues were obtained at about 4 atm total pressure; thus, according to Equation 34 and 36, the total pore volumes computed are for pore regions accessible through openings (D_m) less than about 3 μ. The vitrains used by Toda [117] (420 μ average particle diameter) varied in ash content only from about 1 to 5 mass%, and mercury densities were apparently obtained at 1 atm. The minimum pore diameter penetrated by mercury at this pressure is about 12 μ.

The results shown in Figure 12 indicate that the total pore volumes of the vitrains and whole coals are similar only if the total carbon content is greater than 83%. At lower ranks, the total pore volume for the whole coal is substantially greater than those for the vitrains over corresponding ranges of carbon content. This appears to be the case for coals with a rank of high-volatility B

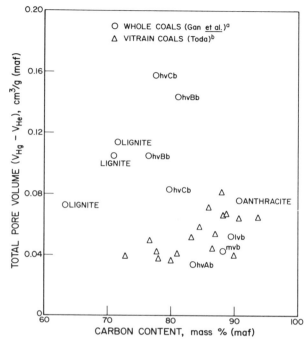

Figure 12 Total pore volumes of coals measured by mercury and helium displacement.

bituminous and lower. Thus it is not possible to accurately predict trends for the lower rank coals solely as a function of carbon content, and a significant fraction of the pore volume of the whole coals probably results from their more heterogeneous composition, compared to vitrains of low-ash content.

Pore volumes in the regions accessible through openings having diameters (D_m) smaller than those corresponding to the results illustrated in Figure 12 were also determined by Gan, Nandi, et al. [97] and by Toda [117]. These results (shown in Figure 13), were based on mercury porosimetry measurements at maximum pressures of 10,000–15,000 psi. In the study by Gan and colleagues, estimates of pore volume at lower values of D_m (12–300 Å) were based on the interpretation of nitrogen adsorption isotherms at 77°K. The results shown in Figure 13 suggest that a substantial portion of the high total pore volumes in whole bituminous coals exists in pores having entry diameters of 12–300 Å.

Variations in total pore volumes during carbonization have been determined by Toda [117] for American and Japanese vitrains. Pore volumes obtained on carbonization at 900°C are compared to original pore volumes in

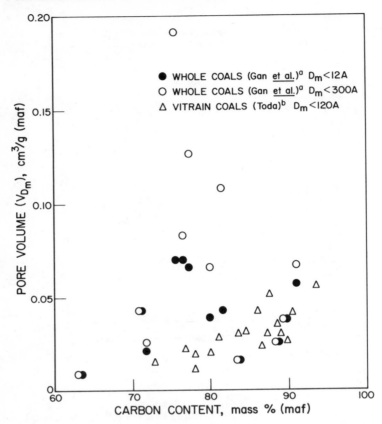

Figure 13 Pore volumes of coals in regions accessible through openings having different entry diameters.

Table 9. These data indicate that char pore volumes, referred to the original coal mass, show an increase compared to the original coal if the initial carbon content is 77.8% or 81.1% but are similar to or smaller than pore volumes of the original coals if the initial carbon content is 86.2%, 87.2%, or 89.8%.

These and other results have been interpreted by Toda to indicate that the pore volumes in macro and transitional pores of caking coals show a net decrease after high-temperature carbonization, due to net particle shrinking. This was not observed for lower-rank noncaking coals. However, Toda found that, at intermediate carbonization stages of caking coals (400–600°C), pore volumes changed drastically, passing through a maximum and then decreasing at higher carbonization temperatures. This was explained by the appearance and disappearance of a "bubble" structure of the caking coal in the plastic stage of carbonization.

TABLE 9 COMPARISON OF COAL AND CARBONIZED CHAR PORE VOLUMES [117]

		Pore Volume (maf)		
			Char	
		Coal		
Initial Carbon Content of Vitrain (maf%)	Char Weight-loss Fraction [g/g coal (maf)]	Coal (cm^3/g)	Char (cm^3/g)	Original Coal (cm^3/g)
77.8	0.41	0.042	0.127	0.075
81.1	0.41	0.041	0.222	0.131
86.2	0.33	0.071	0.063	0.042
87.2	0.24	0.054	0.045	0.034
89.8	0.32	0.039	0.056	0.038

In Toda's study, coals were heated to elevated temperatures at 3°C/min, a relatively low rate, and then maintained at temperature for 15 min. This is of interest because the total pore volume variations that occur with bituminous coals during heat treatment generally must be considered to be partly a function of the time-temperature history, as suggested by Walker and Hippo [99]. With bituminous coals, the evolution of volatile matter during plasticization leads to an initial expansion of individual particles and to the intermediate formation of a bubble structure, reflected in an increase in pore volume of the macro and, to some extent, transitional pores. In Toda's study the exposure of carbonized chars to higher temperatures for extended times apparently led to subsequent particle shrinkage. In other studies, where bituminous coals have been heated to relatively high temperatures at rapid rates (in dilute solid-phase transport or free-fall systems) and then quenched after a few seconds or less, the expanded bubble structure of the resultant chars can be retained. Feldmann, Mima, et al. [119], studying the hydrogasification of bituminous coals at elevated pressure in a free-fall system, found the gross volume of char particles to have increased several-fold, relative to the original coal particle volumes, for particle residence times of a few seconds or less. Thus it is currently recognized that bituminous coal-char particles can have widely different pore structures, depending on the initial heat treatment conditions. Detailed analyses of the complex physical and chemical processes that affect the formation of these structures have only recently been initiated [120].

When precarbonized coal chars are subsequently gasified in reactive atmospheres, total pore volumes generally increase with increasing carbon conversion levels. Some smoothed results obtained for the gasification of chars derived from various coals are shown in Figure 14, and pertinent details of

44 FUNDAMENTALS OF COAL GASIFICATION

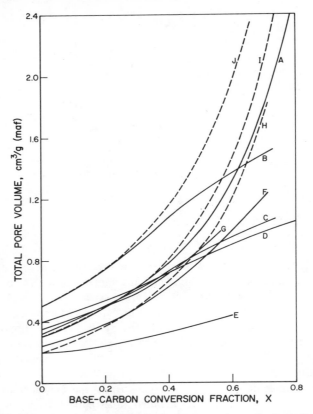

Figure 14 Variations in total pore volumes of coal chars as a function of carbon conversion level during gasification in reactive atmospheres.

the studies represented are given in Table 10. In Figure 14 pore volume variations are given as a function of the base-carbon conversion fraction, X, which represents the fraction of carbon in the initial precarbonized char that is gasified during subsequent treatment in the gasifying medium. For most practical purposes, this conversion term is directly equivalent to other conversion terms applied to coal-char-gasification systems, such as carbon "burn-off" and char weight-loss fraction.

Although most of the curves shown in Figure 14 are approximately similar (taking into account differences in the pore volumes of precarbonized chars), the increases in pore volumes at higher conversions, with one exception (curve A), are significantly less than the volumes originally occupied by the components of the solids that were gasified. This can be illustrated by calculating the pore volume variation that would occur if the external particle volume dimen-

TABLE 10 CONDITIONS FOR STUDIES ILLUSTRATED IN FIGURE 14

Curve (Figure 14)	Reference	Coal Type	Carbonization Conditions Temperature (°C)	Gas	Gasification Conditions Temperature (°C)	Pressure (atm)	Gas
A	121	Lignite	927	N_2	927	35	H_2–H_2O
B	121	High-volatility A bituminous (air-pretreated)	927	N_2	927	35	H_2–H_2O
C	121	Lignite	927	N_2	927	35	H_2
D	121	Subbituminous A	927	N_2	927	35	H_2
E	121	Anthracite	927	N_2	927	35	H_2–H_2O
F	122	High-volatility A bituminous	950	N_2	900	1	CO_2
G	123	Lignite	700	N_2	650	1	H_2O
H	Calculated	$V_p^o = 0.2$	—	—	—	—	—
I	Calculated	$V_p^o = 0.3$	—	—	—	—	—
J	Calculated	$V_p^o = 0.5$	—	—	—	—	—

sions remained constant during gasification and pore volumes increased in direct proportion to the solids gasified. The following expression represents this behavior as a function of initial pore volume, true solids density, and carbon conversion fraction:

$$V_p = \frac{V_p^o + (X/\rho_{He})}{1 - X} \qquad [38]$$

where

V_p^o = total pore volume of carbonized char (in cm^3/g carbonized char)

X = base-carbon conversion fraction of char [in (g carbon gasified/g carbon char)]

ρ_{He} = true density of char in g/cm^3

V_p = total pore volume of gasified char in cm^3/g gasified char

Curves H, I, and J in Figure 14 were calculated from Equation 38 for initial pore volumes of 0.2, 0.3, and 0.5 cm^3/g, respectively, assuming a reasonable true solids density of 2.1 g/cm^3. Although most of the experimental curves are reasonably consistent with the calculated curves up to carbon conversions of 0.4–0.5, experimentally obtained pore volume increases at higher conversions are usually less than the calculated increases, which suggests possible particle shrinkage. The curves depicting Johnson's [121] results (A-E) represent total pore volumes with entry diameters less than 120 μ. Johnson verified particle shrinkage with anthracite, high-volatility bituminous char, and lignite char by direct measurement of the individual particle dimensions before and after gasification in hydrogen at 35 atm and 927°C. In a series of tests, a few particles from each of the different chars were photographed in several orientations under optically calibrated conditions. The particles were then gasified in a thermobalance apparatus to yield carbon conversion fractions of 0.7–0.8 and were photographed again. Detailed examination of the photographs all showed significant reduction in particle volumes consistent with estimates made from mercury displacement measurements. The photographic evidence indicated that the fraction of volume reduction was independent of initial particle diameter in the range 200–800 μ and that external topological characteristics remained unchanged except for a decrease in size. This indicated that the observed shrinkage occurred throughout the individual particles and was not the result of preferential gasification of external char surface layers, nor was it caused by particle attrition or decomposition.

In Johnson's study the rates of gasification of lignite in a steam-hydrogen mixture (pore volume characteristics shown by curve A in Figure 14 were seven

PHYSICS OF GASIFICATION SYSTEMS   47

times greater than the largest gasification rate obtained with the chars shown in curves *B-E*. Since curve *A* was the only curve that showed approximate agreement with calculated pore volume variations for the case of no particle shrinkage, this indicated that dynamic modifications, which could occur within the coal structure during gasification and alter gross dimensions, are inhibited when gasification rates are sufficiently large.

For particles gasified in a fluidized bed rather than a fixed-bed system, such as that used by Johnson, particle attrition could contribute to some extent to decreasing internal particle-void volumes. Results shown in Figure 14 for particles gasified in a fluidized bed (curves *F* and *G*), however, do not exhibit characteristics that suggest significant attrition in addition to particle shrinkage. Dutta, Wen, et al. [124] have reported that coal-char particles gasified in carbon dioxide tend to disintegrate at high levels of carbon conversion fraction (>0.8). Such disintegration of coal-char particles would be favored by high gasification

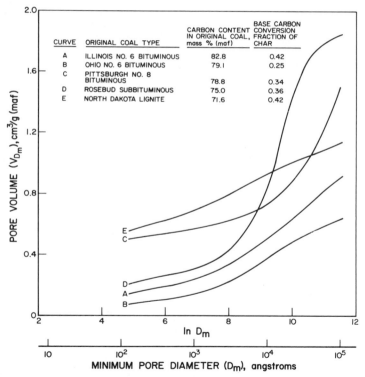

Figure 15 Pore volume distributions of coal chars produced from integral fluid-bed-gasification experiments in process-development studies.

rates, which inhibit particle shrinkage and instead result in highly porous and friable structures at high conversion levels.

Pore volume distributions determined by mercury porosimetry were reported by Stacy and Walker [125] for a variety of coal-char residues resulting from gasification in integral systems, in different process development studies. Some of their smoothed results are shown in Figure 15 for char residues with base-carbon conversions of 0.25–0.42. The results obtained by these investigators, for a total pore diameter range of 130–100,000 Å, indicate that there are relatively small pore volumes for pore diameter openings of about 130–2000 Å, compared to pore volumes for the interval of 2000–100,000 Å. Significant differences in pore volume are apparent for the various chars with pore entry diameters of less than 130 Å, and these differences cannot be accounted for simply by differences in the rank of the coal from which the char was derived.

Pore volume distributions obtained by Johnson [121] for different coal

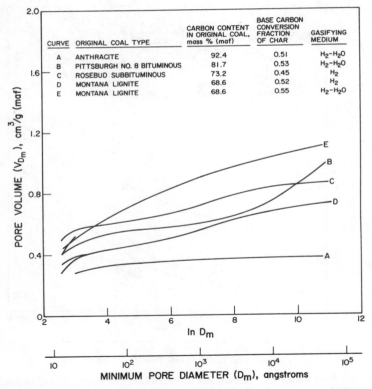

Figure 16 Pore volume distributions of coal chars produced from gasification in a thermobalance apparatus at 327°C and 33 atm. Data from Johnson [121].

chars gasified in hydrogen and in steam–hydrogen mixtures in a thermobalance apparatus are shown in Figure 16. Average particle sizes of the initial coals used in this study were 600 μ. Pore volume evaluations were made by mercury porosimetry for values of D_m greater than 120 Å and by interpretation of nitrogen adsorption isotherms for values of D_m of about 13-120 Å.

The results shown in Figure 16 indicate that for most of the chars, the regions with pore entry diameters of about 55–400 Å contribute little to total pore volume, with major variations in pore volume occurring outside this range. However, the pore volume distribution of lignite that has been rapidly gasified in a steam–hydrogen mixture (curve E) shows that pore volume varies continuously over the total range of pore diameters examined. Total pore volumes obtained with this lignite at different base-carbon conversion levels were described in Figure 14 (curve A). Comparison of curve E with curve D (Figure 16), which illustrate results obtained with the same lignite gasified in steam–hydrogen mixtures and hydrogen alone, indicates that the much more rapid gasification rates observed with steam–hydrogen mixtures contribute to increased pore volumes mainly in the pores having entry diameters greater than 13 Å, that is, in transitional pores and macropores rather than in micropores.

Char pore volumes for values of $D_m < 130$ Å and for the range 130 Å $< D_m < 50{,}000$ Å are represented in Figures 17 and 18 as a function of the rank

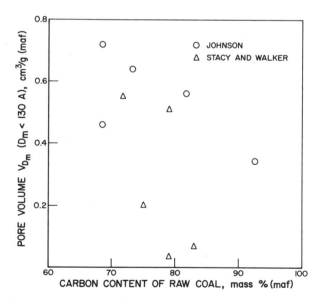

Figure 17 Pore volumes of partially gasified coal chars in regions accessible through openings having diameters less than 130 Å. Data from Johnson [121] and Stacy and Walker [125].

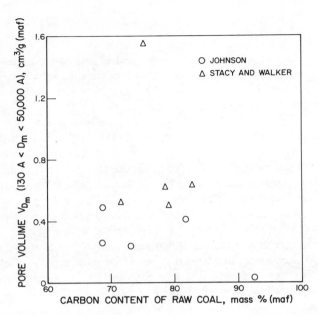

Figure 18   Pore volumes of partially gasified coal chars in regions accessible through openings having diameters of 130-50,000 Å. Data from Johnson [121] and Stacy and Walker [125].

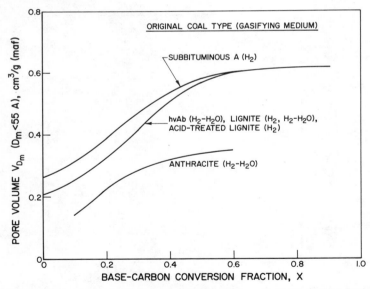

Figure 19   Effect of carbon conversion level on pore volumes of partly gasified chars in regions accessible through openings having diameters less than 55 Å. Data from Johnson [121].

of the coal from which the char was derived, based on results obtained by Stacy and Walker [125] (Figure 15) and by Johnson [121] (Figure 16). In both Figures 17 and 18 there are relatively slight systematic trends in pore volume variations with respect to the carbon content of the raw coals, apparently due to unexplained variances in physical and chemical structures of the raw coals and to differences in the conditions of char formation.

Johnson has found a more systematic relationship between pore volume and carbon conversion values for $D_m < 55$ Å, as shown in Figure 19. On the basis of results such as those shown in that diagram, Johnson has suggested that when gasification rates are not too great, coal-char shrinkage occurs during gasification primarily because of contraction of the microporous volume through crystallite reorientation, as reflected by the lack of a significant increase in micropore volumes at base-carbon conversion fractions greater than about 0.5. He concluded that macropores at these higher levels of conversion also shrink, but in a manner analogous to the shrinkage of cavities in a metallic solid undergoing thermal contraction.

Internal Surface Areas of Coals and Coal Chars. The internal surface area of coals and of carbonized and gasified chars has been measured in a variety of studies to obtain information on variations in structural characteristics during gasification. Although different experimental techniques have been used to estimate the internal surface area of coal-derived solids, the most widely used technique has been analysis of gas-adsorption isotherms obtained in nitrogen or carbon dioxide. Nitrogen isotherms have usually been measured at 77°K and interpreted in terms of the Brunauer, Emmett, and Teller (BET) [126] equation to obtain surface areas. The relative adsorption pressures usually considered in application of the BET equation range 0.05-0.3. Adsorption isotherms obtained in carbon dioxide are usually measured at 298°K, at relative adsorption pressures less than 0.05, and are interpreted in terms of the Dubinin-Polanyi equation [127] to compute surface areas.

Computations of the surface areas of coal-derived solids based on nitrogen and carbon dioxide adsorption isotherms are not often in agreement; for reasons described here in the following text, greater significance is usually attributed to values obtained in carbon dioxide [97, 121, 128-131]. Penetration of nitrogen into the micropore structure of coals or carbonized coal chars is apparently limited by slow, activated-diffusion processes at 77°K, so that the low values of apparent surface areas at most reflect areas of macropores and transitional pores. For partially gasified coal chars, which have more open micropore structures, capillary condensation of nitrogen can lead to unreasonably high values of apparent surface area [132].

Adsorption isotherms obtained with carbon dioxide at the higher temperature of 298°K facilitate activated diffusion into micropore structures, and capil-

52 FUNDAMENTALS OF COAL GASIFICATION

lary condensation is inhibited by the low relative adsorption pressures. There is, however, some question about whether carbon dioxide adsorption isotherms should be interpreted in terms of micropore volume rather than micropore surface area. Although it has been argued that carbon dioxide adsorption on a carbon surface should be limited to a monolayer thickness because of the strong adsorption potential of carbon dioxide molecules with carbon surfaces as well as high intermolecular attraction in the adsorption plane [97, 128, 130], it has also been suggested that, in micropores of molecular dimensions, adsorption forces from opposing surfaces overlap and the concept of monolayer filling of the pores is invalid [133]. Because of this controversy, values for both the surface area and the volume have been reported in the literature, as interpreted from carbon dioxide adsorption isotherms in terms of the Dubinin–Polanyi equation. Both surface area and volume values are, however, convertible to each other from assumed values of the density of the adsorbed carbon dioxide phase and of the equivalent surface coverage area per carbon dioxide molecule. For convenience, the following discussion considers only surface-area values. These are based either on directly reported values or on calculated values obtained by converting reported micropore volumes to equivalent surface areas, assuming an adsorbed carbon dioxide density of 1.03 g/cm^3 and an equivalent surface coverage of 25.3 Å^2/molecule.

Nitrogen surface areas, S_{N_2}, determined by Gan, Nandi et al. [97] for a variety of American whole coals (Figure 20), are relatively negligible for low- or high-rank coals, whereas values up to 93 m^2/g (maf) were obtained for coals in the bituminous range. Carbon dioxide surface areas, S_{CO_2}, for these same coals

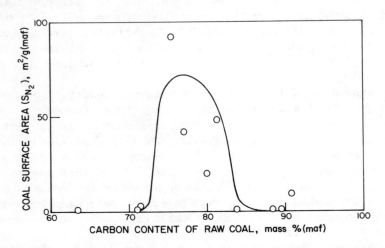

Figure 20 Variation of nitrogen surface areas of coals with carbon content. Data from Gan, Nandi, et al. [97].

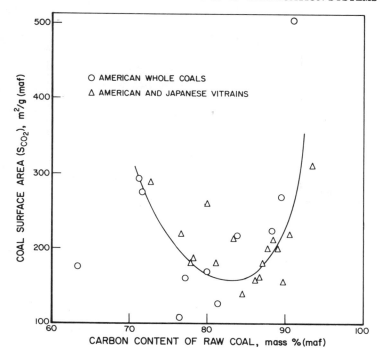

Figure 21 Variation of carbon dioxide surface areas of coals with carbon content. Data from Gan, Nandi, et al. [97].

are shown in Figure 21, along with values of S_{CO_2} computed from Toda's [117] reported micropore volumes (determined by carbon dioxide adsorption) for a variety of American and Japanese vitrains. Values of S_{CO_2} are generally much higher than values of S_{N_2} and tend to follow a reverse trend with respect to the relationship between surface area and rank, compared with values of S_{N_2}. It is of interest that values of S_{CO_2} obtained by Gan, Nandi, et al. [97] and by Toda [117] are relatively consistent over the total range of overlapping rank (about 72-90% carbon in Figure 21), whereas total pore volumes of these same coals and vitrains (Figure 12) are sustantially different for carbon contents less than about 83%. Since almost all of the internal surface area of coals exist in micropores, these results suggest a somewhat systematic relationship between micropore internal surface area and coal rank, in contrast with the significant randomness in macropore and transitional pore structures that can occur in coals of comparable rank, arising from effects of organic and mineral heterogeneities.

Values of S_{CO_2} obtained with carbonized coal chars prepared under comparable conditions are shown in Figure 22 as a function of the carbon content of the raw coals. This plot shows a relatively systematic decrease in S_{CO_2} with

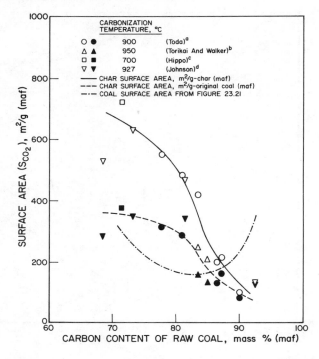

Figure 22 Variation of carbon dioxide surface areas of carbonized coal chars with carbon content of the raw coals. Data from Toda [117], Torikai and Walker [122], Hippo [123], and Johnson [121].

increasing coal rank, compared to the behavior obtained for raw coals (reproduced from Figure 21).

The values of S_{CO_2} for carbonized chars (Figure 22) are referred to both the char mass and the original coal mass. Assuming no attrition or agglomeration during carbonization, the values of S_{CO_2} referred to the coal mass can be considered proportional to the surface area on a per particle basis. On this basis, values for the surface area are closer to those obtained with the raw coal and, in fact, are less than raw coal values for carbon contents greater than about 85%. The overall results described may correspond to a net increase in available surface area during carbonization of coals with carbon contents less than 85%, due to the predominant effect of the removal of volatile matter. For coals having carbon contents greater than 85%, however, less volatile matter is evolved; and the net effect of carbonization under these conditions is a decrease in surface area, due possibly to shrinkage or closure of pore-opening regions induced by the heat treatment.

In some studies coal-char surface-area variations during the course of gasification in reactive atmospheres have been determined as a function of the carbon conversion level. Results obtained by Hippo [123] for the gasification of Montana lignite in steam at 1 atm are shown in Figure 23. A fluidized-bed reactor was used in this study, and chars were initially prepared in nitrogen. The results shown in Figure 23 indicate an initial increase in surface area (S_{CO_2}) for carbon conversions less than 10–20%, with a relatively slight decline in surface area thereafter for carbon conversions up to 56%. Hippo has suggested that the increase in surface area is due to a reduction in pore restrictions during initial gasification stages, but that the pore walls collapse in later stages, leading to the joining of pores and the resultant decrease in surface areas.

Torikai and Walker [122] studied surface area variations of coal chars that were prepared by carbonization of hvbA coals and were then gasified in carbon dioxide in a fluidized-bed reactor at one atmosphere. In this study two coal lithotypes obtained from the same seam were used: (1) a relatively low-vitrinite material classified as a durain and (2) a high-vitrinite material classified as a clarain. The raw coals were first pretreated in air at a low temperature to reduce agglomeration tendencies during the subsequent carbonization. Surface-area variations (S_{CO_2}) of partially gasified chars obtained with these samples are shown in Figure 24. The trends indicated are qualitatively similar to those obtained by Hippo, with surface areas initially increasing with increasing carbon conversion, reaching a maximum at intermediate carbon conversion and then

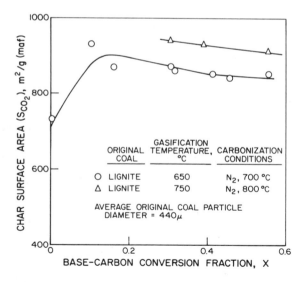

Figure 23 Effect of carbon conversion level on carbon dioxide surface area of lignite coal chars gasified in a fluid bed at 1 atm with steam. Data from Hippo [123].

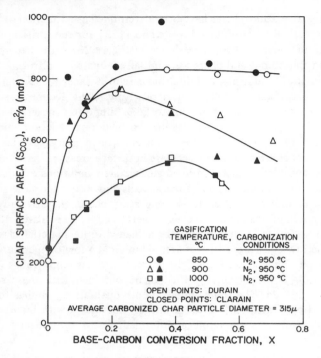

Figure 24 Effect of carbon conversion level on carbon dioxide areas of bituminous coal chars gasified in a fluid-bed atmosphere with carbon dioxide. Data from Torikai and Walker [122].

decreasing with further increases in carbon conversion. Maximum surface areas increase with decreasing gasification temperature from 1000°C to 850°C, although the variation is relatively slight from 900°C to 850°C. At still lower temperatures, it would be expected that this trend would reverse for chars carbonized at gasification temperatures because of the effect of residual volatile matter in inhibiting gas penetration into pores during the measurement of adsorption isotherms.

Such inhibition may account for the somewhat higher apparent surface areas obtained by Hippo for lignite chars resulting from gasification at 750°C (carbonization temperature = 800°C), compared to surface areas obtained for lignite chars resulting from gasification at 650°C (carbonization temperature = 700°C). The decrease in surface area at higher carbon conversions shown by the data of Torikai and Walker is relatively slight at the gasification temperature of 850°C, but quite severe at the gasification temperature of 900°C, indicating the sensitivity of char structural characteristics to temperature.

Although Torikai and Walker obtained similar results with chars from two different lithotypes of the same rank, surface-area characteristics during gasification would be expected to depend on initial coal properties as well as conditions of carbonization and gasification.

Johnson [121] studied surface-area variations for a variety of coal chars gasified in hydrogen and steam–hydrogen mixtures, in a thermobalance apparatus, at 35 atm. Johnson's results, shown in Figure 25, showed only slight decreases in surface areas during gasification at 927°C for chars derived from anthracite, high-volatile bituminous coal, and lignite. Only with char obtained from a subbituminous A coal did the surface areas decrease with carbon conversion to an extent comparable to that observed by Torikai and Walker for bituminous char gasified at 900°C. In Johnson's study, similar surface-area trends were obtained for the gasification of individual chars in both hydrogen and steam–hydrogen mixtures. In addition, no significant effects on lignite char surface-area characteristics were observed when initial acid pretreatment was employed, in spite of the fact that chars derived from acid-treated lignite were

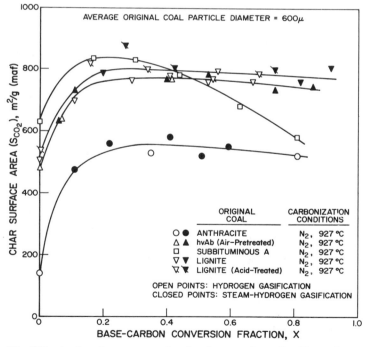

Figure 25 Effect of carbon conversion level on carbon dioxide surface areas of different coal chars gasified in a thermobalance apparatus at 927°C, 35 atm, in hydrogen and steam–hydrogen mixtures. Data from Johnson [121].

58 FUNDAMENTALS OF COAL GASIFICATION

much less reactive than chars derived from untreated lignite. Significantly, Johnson also found essentially identical char surface-area characteristics during the gasification of chars formed by initial carbonization in nitrogen and chars formed directly in the reactive-gas atmosphere.

The maximum surface areas exhibited by the curves drawn in Figures 23-25 for chars gasified at temperatures of 650-927°C are plotted in Figure 26 as a function of the carbon content of the raw coal. Considering the variations in gasification conditions and original coal types used in these studies, the maximum char surface areas are surprisingly similar for original coal carbon contents of 68-85%. The anthracite char surface area, however, is significantly less than the area of lower-rank chars. The dashed curve in Figure 26 represents a correlation proposed by Johnson [121] to define relative gasification reactivities of

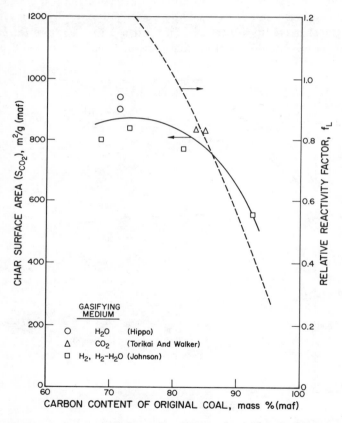

Figure 26 Variation of maximum carbon dioxide surface areas of coal chars with carbon content of the original coals. Data from Hippo [123], Torikai and Walker [122], and Johnson [121].

coal chars as a function of original coal carbon content (see Figure 50). Although there appears to be a possible relationship between variations in char surface areas and variations in gasification reactivities for chars derived from coals having carbon contents greater than about 82%, chars derived from lower-rank coals exhibit higher reactivities than can be attributed solely to variations in surface areas.

KINETICS OF GASIFICATION REACTIONS

Rational design of coal gasifiers requires the best possible understanding of fundamental chemical and physical processes that occur: (1) during the transition of coal to coal char and (2) during the subsequent gasification of the coal char itself. The following sections focus primarily on studies relevant to these two main reaction stages, in which attempts have been made to eliminate or separately account for mass and heat-transfer effects with the hope of characterizing intrinsic chemical behavior in the experimental gasification systems employed. It should be kept in mind, however, that it is not often evident to what extent this objective has been achieved in individual studies and that, as suggested by Ergun and Mentser [134], it is likely that some of the wide variation in kinetic data reported in the literature is due in large measure to differences in experimental systems and to difficulties in making a quantitatively complete evaluation of physical factors.

Initial-State Coal-Gasification Kinetics

Gaseous hydrocarbons produced during the initial coal-gasification stages are extremely important in processes to convert coal to substitute natural gas. Significantly, when raw coals are heated in an atmosphere containing hydrogen at elevated pressure, greatly increased gaseous hydrocarbon yields are obtained, compared to those obtained when devolatilization occurs in an inert-gas atmosphere. However, this high reactivity is transient, lasting only for seconds or less at elevated temperatures corresponding to the time required for conversion of the coal to the relatively much less reactive char.

Because of its importance, this phenomenon, first observed by Dent [135] in the 1940s, has been studied in a variety of experimental investigations, particularly during the last 15 years. In some of the earliest work during this period, it was usually believed that these excessive gaseous hydrocarbon yields, particularly of methane, resulted from hydrogenation or hydrogenolysis of the evolved volatile matter, and not from heterogeneous interactions involving gaseous hydrogen and the coal itself. This idea was proposed by Birch, Hall, et al. [136] and by Blackwood and McCarthy [137] for the gasification of a brown coal in a fluidized bed and later by Feldkirchner and Linden [138] as well as by Wen and Huebler [139] for the gasification of bituminous coal in a fixed bed. However, Moseley and Patterson [140, 141] were the first investigators to propose that although excessive methane yields resulted in part from hydrogenation of volatile matter, a significant fraction did indeed result from heterogeneous coal-hydrogen reaction. Subsequently, other investigators have also usually agreed that heterogeneous interactions play an important role in initial methane formation reactions. There is, however, considerable variation in the mechanisms proposed and the kinetic representations made to explain experimentally observed behavior, a diversity of opinion that is understandable in view of the complexity of the reaction processes that can occur during initial coal-gasification stages, as well as the difficulty in accurately characterizing experimental parameters during extremely short reaction times.

Some of the overall reactions that can occur during the initial coal gasification stages are simply described in Figure 27. In an inert-gas atmosphere the devolatilization reactions represented are similar to the pyrolysis model proposed by van Krevelen [37] whereby coal pyrolysis is divided into three overall processes. Initially, the coal undergoes a kind of depolymerization reaction, which leads to the formation of a metastable intermediate product. The product then undergoes cracking and recondensation to result in the evolution of primary gases and oils, yielding semichar. After the rate of devolatilization has

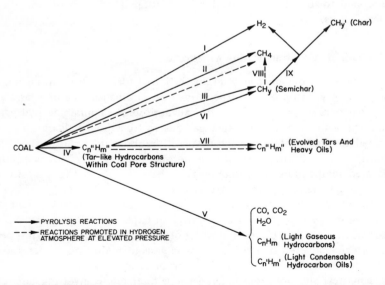

Figure 27a Simplified reaction scheme. Direct coal reactions during initial gasification stages.

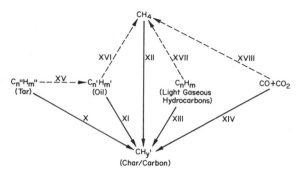

Figure 27b Simplified reaction scheme. Secondary reactions.

passed a maximum, the third reaction becomes important, in which the semichar is converted to char primarily through the evolution of hydrogen.

In Figure 27a, primary devolatilization in an inert atmosphere is illustrated by reactions I-VIII, which involve the evolution of coal oxygen as carbon oxides and water, in addition to the formation of other gaseous hydrocarbons, heavy oils and tar, and some gaseous hydrogen. The solid product of this reaction stage is semichar. Secondary devolatilization is illustrated by reaction IX, in which semichar is converted to char through the evolution of hydrogen. In a hydrogen atmosphere at elevated pressure, (reaction VII) higher yields of evolved oils and tars are promoted over the internal cracking of oils and tars (reaction VI). Methane formation from the direct reaction of gaseous hydrogen and coal is indicated by reaction II, which occurs concurrently with primary devolatilization reactions. Reaction VIII, on the other hand, illustrates another path for the formation of methane in hydrogen, which results from heterogeneous interaction during secondary devolatilization concurrent with conversion of semichar to char, along with associated hydrogen evolution.

Figure 27b illustrates the secondary reactions of products formed from the direct coal reactions. In an inert atmosphere, the various hydrocarbon products and carbon oxides primarily are cracked to form carbon or char (reactions X-XIV). In a hydrogen atmosphere, however, initial hydrogenation products can undergo hydrogen enrichment, leading to the formation of light oils, gaseous hydrocarbons, and eventually methane, as shown in reactions XV-XVII. The possibility of methanation of carbon oxides is also included, as in reaction XVIII.

In terms of the reaction scheme shown in Figure 27, excessive methane or other light gaseous hydrocarbon yields can result during the initial coal-gasification stages in a hydrogen atmosphere from: (1) direct hydrogenation of coal or semichar (reactions II and VIII), which also should result in an increase in the amount of coal carbon that is gasified or (2) from the hydrogenation of other gaseous hydrocarbons, oils, tars, and carbon oxides, which, for the most

TABLE 11 NOMINAL FOR INVESTIGATIONS OF INITIAL COAL-GASIFICATION REACTIONS IN HYDROGEN-CONTAINING GASES

Reference	Coal Type	Primary Particle Diameter Range (μ)	Feed Gases	Temperature Range (°C)	Pressure Range (atm)	Residence Time at Temperature (sec) Solids	Residence Time at Temperature (sec) Gas	Solids Heat-up Rate (°C/sec)
		Fixed-Bed Studies						
142	Low-temperature bituminous coal char	420–1190	H_2, H_2O, H_2–H_2O	900–1150	70	0–1500	—	~20
143	Pittsburgh high-volatility bituminous coal	250–590	H_2	800, 1200	5–70	0–1800	<1	~10
7	Pittsburgh high-volatility bituminous (air-preheated) coal	420–840	N_2, H_2, H_2O, H_2–H_2O, H_2–CH_4	450–900	10–70	60–3600	—	~10
144–146	Pittsburgh high-volatility bituminous coal, Montana lignite	50–80	H_2, He, N_2, H_2–He	400–1100	0.001–71	0–20	—	60–10^4
147, 148	Illinois high-volatility bituminous coal	<40	H_2, N_2	650–1000	100	2–65	0.2–23	20–1400
149	North Dakota lignite	40	H_2	550–850	10–280	0.3–10	0.3–10	~10^4
		Fluidized-Bed Study						
136	Brown coal	300–760	H_2, H_2–N_2, synthesis gas	500–950	2–40	900–6500	6	—

Dilute Solid-phase Transport Studies

141	British high-volatility bituminous coal	100–150	H_2	840–930	50–500	0.5–5	60–120	—
150	Pittsburgh high-volatility bituminous coal	150–300	H_2–CH_4	730	100–200	1–2	120–240	—
151	North Dakota lignite	70–150	Synthesis gas	940–970	70–85	2–10	2–10	—
119 152	Pittsburgh high-volatility bituminous coal, Illinois high-volatility bituminous coal, North Dakota lignite	70–300	H_2, H_2–CH_4	650–900	35–140	1–2	30–200	—
153– 155	Montana, lignite, Montana subbituminous coal, North Dakota lignite	70–90	H_2, He, H_2–He	450–850	20–50	5–15	5–15	30~10^4
156	North Dakota lignite, New Mexico subbituminous coal	<50, <150	H_2, A	700	70–100	15–40	10–70	—

part, would not directly contribute to an increase in the total amount of coal carbon gasified. One exception would be through a combination of reactions VII, XV, and XVI, which would increase the total amount of carbon gasified, as well as the methane yield.

Various investigations have been conducted to study initial reactions during coal gasification in hydrogen-containing gases. The range of conditions employed in some of these investigations is given in Table 11. Three types of experimental systems were used in these studies: (1) fixed coal bed-flowing gas, (2) fluidized coal bed, and (3) dilute solid-phase transport, or free-fall systems. In addition to differences in the coal type and the range of conditions employed, there are also differences in the methods for characterizing experimental results. In some of the fixed-bed studies, gasification behavior was characterized on the basis of total coal weight loss or coal–carbon conversion [7, 144–146]. In such cases, although little information is directly available describing gasified produce yields or secondary reactions, the data do reflect the net gas–solid interactions leading to enhanced carbon gasification rates during initial reaction stages. In other studies, more complete information has been obtained, including yields of major gasified species or species groups, thus providing a basis for interpretation of individual reactions that occur.

In one of the first major studies since Dent's work, Birch, Hall, et al. [136] investigated the gasification of brown coal with hydrogen in a continuous fluidized-bed system at elevated pressures. Results showed a substantial increase in methane yield with increasing hydrogen pressure during the initial-stage reactions, as well as during the much slower char–hydrogen reaction, which reaction was significant at the relatively long char residence times employed.

The authors described the gasification process as occurring in two fairly well-defined stages, an initial rapid stage associated with devolatilization processes and a second, much slower stage associated with the char–hydrogen reaction. The first stage was considered to be essentially instantaneous, relative to the amount of time required for significant char–hydrogen reaction. This is because the first-stage reaction involves rapid reaction of hydrogen with the more reactive parts of the coal structure such as oxygen-containing functional groups and aliphatic hydrocarbon side chains.

It was found that although methane yields tend to increase with increasing hydrogen partial pressure, the maximum yield during the initial stage is stoichiometrically limited and is independent of hydrogen pressure. For the brown coal used, this maximum yield was estimated at about 28% conversion of the carbon feed to methane and 12% to other volatiles. The authors pointed out that the amount of solid carbon remaining after completion of the first reaction stage corresponds fairly closely to that known to be present in aromatic ring structures in brown coal.

A similar explanation was postulated by Blackwood and McCarthy [137], who later analyzed the data of Birch, Hall, et al. [136] in greater detail, using an extrapolation procedure to distinguish between first-stage and second-stage methane yields. Blackwood and McCarthy concluded that, at temperatures above 700°C, methane yields obtained during the first-stage reaction are dependent only on hydrogen partial pressure up to the maximum yield possible, which was obtained at a hydrogen partial pressure of 30–40 atm (see Figure 28).

The model proposed by Blackwood and McCarthy suggests that there is a fixed fraction of reactive carbon in the raw coal structure that can potentially be converted to methane, once sufficient time has elapsed to complete devolatilization processes. During the main devolatilization process the hydrogenation reactions leading to methane formation compete with certain of the pyrolysis processes that tend to form carbon oxides, heavier hydrocarbons, and char or fixed carbon. Within the context of the simplified reaction scheme shown in Figure 27a, high-pressure hydrogen tends to favor reaction II in competition with reactions IV and V.

A two-stage gasification model has also been proposed by Wen and Huebler [139] in analyzing data obtained by Feldkirchner and Huebler [142] for the gasification of lignite and a low-temperature bituminous coal char in hydrogen. Based on the improved time resolution of the data obtained by Feldkirchner and

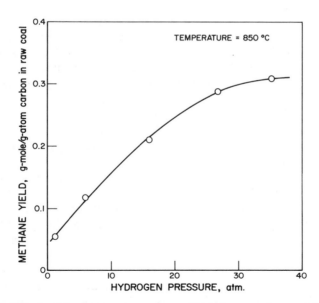

Figure 28 Effect of hydrogen pressure on first-stage methane yields obtained for gasification of brown coal in a fluid bed. Data from Blackwood and McCarthy [137].

66 FUNDAMENTALS OF COAL GASIFICATION

Huebler with a fixed-bed flowing gas reactor and differential gas conversion, Wen and Huebler proposed the following correlation to account for methane yields obtained during the first-stage reaction in hydrogen:

$$\frac{dY_{CH_4}}{dt} = k(f - X)P_{H_2} \qquad [39]$$

where

X = total carbon gasified (in g/g of feed carbon)
Y_{CH_4} = methane yield (in g/g of feed carbon)
f = fraction of feed carbon that can react by first-stage reactions
P_{H_2} = hydrogen partial pressure (in atm)
t = time (in sec)
k = rate constant (5.6×10^{-4} atm^{-1} sec^{-1})

The variable f is a function of coal type and varies slightly with temperature at 800-950°C.

In the correlation of Equation 39, Wen and Huebler suggested values at 927°C of f = 0.25 for low-temperature bituminous coal char and f = 0.48 for lignite. It should be noted that the term $f - X$ represents the fraction of potentially volatile carbon still remaining in the reacting coal particle at any time. Thus Wen and Huebler's model proposes that methane yields obtained during the first-stage reaction result from direct interaction of hydrogen with reactive groupings remaining in the coal, a process that competes with pyrolysis reactions involving these groups. In this sense the model is essentially identical to that proposed by Blackwood and McCarthy, although Wen and Huebler's formulation indicates that only the gasified product yield is affected by gaseous hydrogen and not the total amount of carbon gasified during the first-stage reaction.

Moseley and Patterson [140, 141] concluded that the high yields of methane obtained during the initial gasification stages in hydrogen resulted not only from the hydrogenation of a limited amount of coal volatile matter, or potentially volatile matter, but also from direct hydrogasification of an active intermediate that would otherwise form relatively unreactive char or fixed carbon. These conclusions were based on results obtained in dilute-phase transport reactor systems for the gasification of low-temperature bituminous char and noncaking, high-volatility bituminous coal, with feed hydrogen pressures ranging up to 500 atm.

Results such as those shown in Figure 29 [137], were obtained with a time resolution (0.5-5 sec) greatly improved over that existing in previous studies, showed that at sufficiently high hydrogen pressures, substantial fractions of potential fixed carbon can be gasified in a few seconds. At an average product

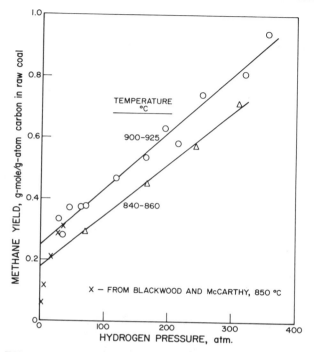

Figure 29 Effect of hydrogen pressure on methane yields obtained from gasification of noncaking bituminous coal in dilute solid-phase transport reactor. Data from Moseley and Patterson [140, 141]; "X" from Blackwood and McCarthy [137].

gas hydrogen partial pressure of about 350 atm with a feed hydrogen pressure of 500 atm, almost complete conversion of the feed-coal carbon was achieved at 900-925°C.

Moseley and Patterson postulated that at elevated temperatures, the hydrogenation of evolved volatile matter to form methane is essentially complete at a hydrogen pressure of less than 100 atm. At higher pressures, methane yield increases linearly with hydrogen partial pressure due solely to the gasification of potential fixed carbon in the intermediate coal structure. According to this description, the lines drawn in plots like Figure 29 should curve downward at low hydrogen partial pressures due to incomplete hydrogenation of volatile matter. A possible example of this effect is shown in Figure 29 by Blackwood and McCarthy's [137] representation of methane yields obtained during the first-stage reaction in Birch's study, at a hydrogen partial pressure of less than 40 atm. Although Blackwood and McCarthy interpreted the indicated trends as corresponding to a leveling off of methane yields with increasing partial pressure, the data of Moseley and Patterson show that yields above around 50 atm are approximately proportional to hydrogen partial pressure.

Moseley and Patterson have suggested that when the hydrogen reaction with volatile matter is complete, methane formation is due solely to the attack by hydrogen on transient, unstable structures in the pyrolyzing coal char, and that this reaction competes effectively with normal thermal stabilization processes leading to unreactive carbon residues. In a formalized model these investigators have proposed the following overall reaction:

$$C_0 \text{ (coal)} \xrightarrow{K_A} V \text{ (volatiles)} + I \text{ (active intermediate)} \xrightarrow{K_B} \text{Cross-linked, less active structure, not easily gasified by hydrogen}$$

$$+ 2nH_2 \downarrow K$$

$$I + nCH_4$$

[4]

where K, K_A, and K_B are rate constants.

The initial reaction corresponds to a purely thermal decomposition–polymerization process that is not influenced by hydrogen pressure and that leads to the formation of volatiles and an active intermediate in the char structure. The active intermediate promotes rapid hydrogenation of the coal-char structure to form methane in competition with a rapid deactivation reaction leading to the formation of unreactive cross-linked structures, with associated loss of reactive intermediates. The formation of reactive intermediates is considered to be rate-limiting, and it is assumed that no net comsumption of the active intermediate occurs in the hydrogenation step. With this model, and assuming first-order reaction processes, the following rate equation was derived to describe the rate of methane formation under isothermal conditions:

$$\frac{dY_{CH_4}}{dt} = \frac{KK_A C_0 \, e^{-K_A t}}{K_B} P_{H_2} \cdot M \qquad [40]$$

where

C_0 = initial concentration of structures in the coal capable of forming the active intermediate

M = effective char concentration

Moseley and Patterson suggest that the model in Equation 40 represents a drastically simplified picture of a complex process and that not only will C_0,

K_A, and K_B depend on the original coal and the temperature of hydrogenation, but also that K_A and K_B are unlikely to remain strictly constant at any one temperature, since devolatilization and linking require larger activation energies to overcome the hindrance of previous cross-linking. The solid carbon concentration, M, in Equation 40 is not considered to change significantly with time compared to changes in the concentration of the active intermediate. This is equivalent to assuming that the solid carbon is always in excess kinetically even when a large portion of it has been gasified. Although this assumption is consistent with the straight lines drawn in Figure 29, it would be reasonable to expect increasing downward curvature at levels of carbon conversion approaching unity.

The model proposed by Moseley and Patterson, then, suggests that the formation of methane during the initial gasification of coal in hydrogen results in part from hydrogenation of volatile matter—probably through a secondary reaction such as that described in Figure 27b, but also through heterogeneous hydrogenation of the potential fixed carbon in the devolatilizing coal structure. The fact that they suggest that the main agents of activation for the heterogeneous hydrogenation may be groups that normally yield hydrogen under pyrolysis conditions indicates that reaction VIII in Figure 27a occurs instead of reaction II. Since secondary-stage coal–hydrogen evolution is usually thought to occur after primary devolatilization reactions leading to semichar [37], the results of Moseley and Patterson can be interpreted as concluding that the rapid-rate heterogeneous reaction by methane formation also occurs after primary devolatilization.

Zahradnik and Glenn [157] later proposed a model somewhat similar to that of Moseley and Patterson, based on analysis of experimental studies conducted at Bituminous Coal Research, Inc. In this model, primary devolatilization reactions are considered to go to completion very rapidly—in milliseconds at temperatures above 950°C—prior to any significant reaction between gaseous hydrogen and active intermediates formed from the primary devolatilization. Active intermediates then can either react with hydrogen to form methane or deactivate to form inactive char. These processes are shown schematically as follows:

$$\text{Coal} \longrightarrow C^{**} + CH_4 + (CO, CO_2, H_2O, \text{etc.}) \qquad [5]$$

$$C^{**} \xrightarrow{k_1} C^* \qquad [6]$$

$$C^{**} \xrightarrow[H_2]{k_2} CH_4 \qquad [7]$$

70 FUNDAMENTALS OF COAL GASIFICATION

where

C^{**} = active intermediate produced directly from coal when coal is thermally depolymerized

C^* = inactive char produced by reaction of the active intermediate with itself

k_1, k_2 = rate constants

One of the main characteristics of the model is that the active intermediate is consumed in the methane-formation step, contrary to Moseley and Patterson's assumption that there is no net consumption of the active intermediate in this reaction step. Zahradnik and Glenn developed the following correlation, based on their model, to describe total methane yields after complete conversion of the active intermediate either to methane or to inactive char:

$$Y_{CH_4} = \frac{Y^0_{CH_4} + bP_{H_2}}{1 + bP_{H_2}} \quad [41]$$

where

$Y^0_{CH_4}$ = methane yield resulting only from devolatilization

b = kinetic parameter, proportional to the ratio of rate constants for methanation and deactivation of the active intermediate

Kinetic parameters in the preceding reactions and equations were determined to describe experimental results obtained for the gasification in hydrogen of several coals in dilute-phase transport reactor systems. Among the results to be described are certain of the results obtained by Moseley and Patterson. Examples of the values obtained are given in Table 12.

TABLE 12 KINETIC PARAMETERS BY ZAHRADNIK AND GLENN [157]

Coal	Reference	Temperature, (°C)	$Y^0_{CH_4}$	b (atm^{-1})
North Dakota lignite	151	950	0.07	0.0067
Elkol subbituminous	158	950	0.08	0.0060
High-volatility bituminous	141	900–925	0.08	0.0060
High-volatility bituminous	141	850	0.07	0.0035
High-volatility bituminous	150	725	0.09	0.0017

Interestingly, the range of methane yields from pyrolysis and hydrogenation of volatiles given here is considerably lower than the corresponding yields of 0.2-0.25 obtained by Moseley and Patterson for the hydrogenation of volatile matter. This difference apparently arises because Moseley and Patterson used a linear extrapolation with hydrogen pressure to evaluate these yields, whereas Zahradnik and Glenn's model accounts for the downward curvature of yields with decreasing hydrogen pressure. Assuming that the parameter b, equal to the ratio k_2/k_1, is dependent only on temperature and not on coal type, Zahradnik and Glenn determined an activation energy of approximately 15 kcal/g-mol, representing the difference in activation energies between methanation and the deactivation process.

One particularly important aspect of the models and correlations proposed by Moseley and Patterson, as well as by Zahradnik and Glenn, is that they successfully use only the hydrogen partial pressure rather than the total gas composition to characterize the experimental data. From Moseley and Patterson's data and correlations, this implies that the methane partial pressure does not have a significant effect on reaction kinetics for methane concentrations of up to 80% in the product gases. Similarly, for the Bituminous Coal Research, Inc. data, obtained at lower hydrogen partial pressures, the significant concentrations of carbon monoxide, carbon dioxide, and methane were not considered to affect hydrogenation reactivity.

The initial gasification stages of coal were also studied by Johnson [7] and Pyrcioch, Feldkirchner, et al. [159] using a thermobalance apparatus, to measure reaction behavior in terms of the gasification rates of individual coal components under conditions of differential gas conversion. A typical result with this type of experimental system that illustrates the high reactivity exhibited by coals during initial reaction stages in hydrogen is shown in Figure 30. An important effect of using this thermobalance apparatus is that after the coal sample is lowered into the reactor environment, the initial reactions involving devolatilization and methane formation are completed within the first few minutes required for the sample to heat up to reactor temperature. Figure 30 illustrates several results. Curve A shows that for simple pyrolysis conditions, evolution of volatile matter is completed within the first few minutes and the total weight loss is about equal to the volatile matter of the feed coal, as determined by proximate analysis. When the coal-char residue obtained after pyrolysis is exposed to hydrogen at elevated pressures (curve B), relatively slow gasification to form methane occurs. When, however, the feed coal is directly exposed to a hydrogen atmosphere without prior devolatilization in nitrogen (curve C), the total weight loss during the first few minutes significantly exceeds the weight loss obtained under pyrolysis conditions. This excessive weight loss corresponds to the additional carbon gasified primarily to "rapid-rate" methane, with yields increasing with increasing hydrogen pressure (curve D).

72 FUNDAMENTALS OF COAL GASIFICATION

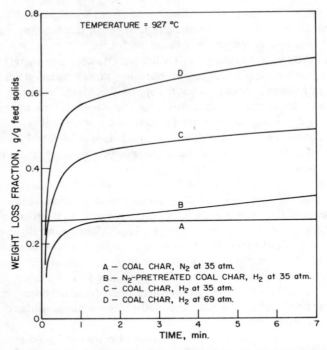

Figure 30 Typical weight losses obtained for gasification of air-pretreated high-volatility bituminous coal in thermobalance. Data from Johnson [7].

Johnson correlated the excessive carbon conversion due to rapid-rate methane formation at final reaction temperatures above 800°C (hydrogen partial pressures up to 70 atm) with the following expression:

$$X = 1 - \exp(-f_R \, 0.0092 P_{H_2}) \qquad [42]$$

where

X = base carbon gasified per base carbon in feed coal;

f_R = reactivity factor used to characterize coal type ($f_R \sim 1.0$ for high-volatility bituminous coals)

P_{H_2} = hydrogen partial pressures (in atm)

(Base carbon is defined as the carbon in the "fixed-carbon" fraction, determined by standard proximate analyses.) As with correlations developed by other investigators, Equation 42 was successfully applied not only to systems using pure hydrogen, but also to systems with hydrogen-methane mixtures and hydrogen-steam mixtures; this implies that the rapid-rate phenomenon occurs only in the presence of gases containing hydrogen and is not affected by other gaseous species.

Equation 42 was developed to describe yields obtained above 800°C, at which temperature devolatilization and those processes leading to rapid-rate methane formation are completed in relatively short times. For gasification at lower temperatures, Johnson proposed a more detailed model and correlation to describe methane yields. The model proposed was similar to that of Moseley and Patterson, but it included the formation of active intermediates, which occurs by an infinite set of parallel reactions having different activation energies. With this assumption, the correlation proposed by Moseley and Patterson was modified by replacing the term $K_A e^{-K_A t}$ in Equation 40 by the following term:

$$\int_0^\infty K_A e^{-K_A t} f(E) dE \qquad [43]$$

where

$$K_A = K_A^0 e^{-E/RT} \qquad [44]$$

The distribution function $f(E)$ is expressed by

$$f(E) = \frac{4\alpha^{3/2}(E - E_1)^2}{\sqrt{\pi}} \exp\left[-\alpha(E - E_1)^2\right] \text{ for } E \geqslant E_1 \qquad [45]$$

and

$$f(E) = 0 \text{ for } E < E_1 \qquad [46]$$

With this formulation, results such as those shown in Figures 31 and 32 [159] were described with the following evaluation of kinetic parameters:

$$E_1 = 18.2 \text{ kcal/g-mol} \qquad [47]$$

$$\alpha = 0.111 \text{ (kcal/g-mol)}^{-2} \qquad [48]$$

$$K_A^0 = 150 \text{ sec}^{-1} \qquad [49]$$

$$\frac{C_0 K}{K_B} = 0.0092 f_R, \text{ atm}^{-1} \qquad [50]$$

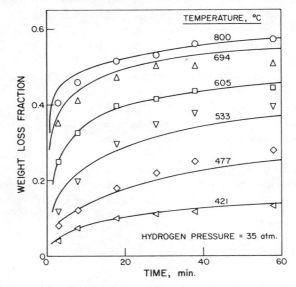

Figure 31 Effect of temperature and residence time on weight loss obtained for hydrogen gasification of air-pretreated high-volatility bituminous coal in thermo-balance. Data from Pyrioch, Feldkirchner, et al. [159].

In describing these curves, additional correlations were developed to define weight-loss contributions due only to devolatilization reactions. The preceding evaluations were based on data obtained for coal residence times greater than 180 sec and should not be extrapolated to shorter residence times, where they are not applicable.

A study of the gasification of lignite and bituminous coal at short residence times ($\leqslant$ 20 sec) has been reported by Anthony and colleagues [144-146] using total coal weight loss as a measure of kinetic behavior. In this study small samples of coal were sandwiched in a steel-mesh screen that could be rapidly heated to a desired temperature and held there for a desired time. The following parameters were measured: coal weight loss, as a function of heating rate (65-100,000°C/sec), coal residence times at temperature (0-20 sec), final temperature (400-1100°C), total pressure (0.001-70 atm), hydrogen partial pressure (0-70 atm.), and coal particle size (70-1000 μ). Tests were conducted using hydrogen, helium, and hydrogen-helium mixtures. It was observed that under these conditions total weight loss increases significantly with increasing hydrogen partial pressure above 10 atm and with increasing temperature, but only slightly with increasing heat-up rate. At temperatures above 990°C, maximum weight loss occurred in 2-3 sec with no significant further weight loss up to 20-30 sec.

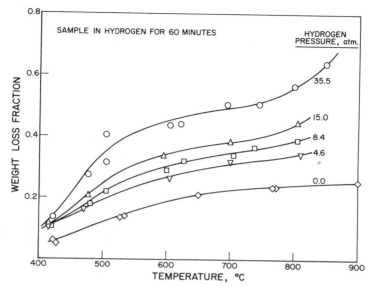

Figure 32 Effect of pressure and temperature on gasification of air-pretreated high-volatility bituminous coal in thermobalance. Data from Pyrcioch, Feldkirchner, et al. [159].

With bituminous coal, at a final temperature of 1000°C and residence times of 5-20 sec, weight losses in helium decreased from 54-37% of the original amount of coal, in a range of pressures of 0.001-70 atm (Figure 33) [144]. Weight losses in hydrogen reached a minimum of approximately 48-50% of the original coal weight in the pressure range 1-10 atm and then rose with further increases in pressure to about 60% of the original coal weight, at 70 atm. Significantly, at 1000°C these variations in total weight loss are due almost solely to variations in the amount of carbon gasified, since the evolution of other coal components—oxygen, hydrogen, and other elements—would be essentially complete.

In explaining these and other results obtained in this study, Anthony, Howard, et al. [145] proposed that two types of volatiles are formed within coal particles due to thermal activation processes: (1) nonreactive species and (2) species that are extremely reactive, particularly in the initial stages of decomposition, when free radicals are likely to exist. The nonreactive volatiles are all evolved from the particle, but the reactive volatiles either evolve directly from the particle, are stabilized and rendered inactive by reaction with gaseous hydrogen or they evolve from the particle, or polymerize or crack on hot coal surfaces, thereby decreasing the volatile yields. The authors further postulated that gaseous hydrogen can also react directly with the coal for a transient period concurrent with the formation of reactive volatiles to produce (presumably) stable,

76 FUNDAMENTALS OF COAL GASIFICATION

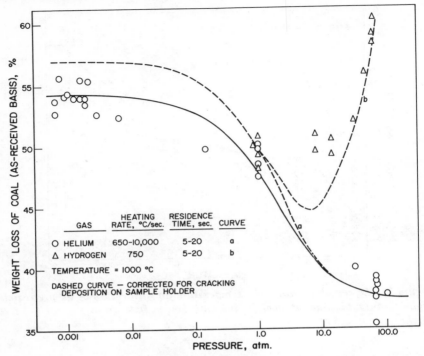

Figure 33 Effect of pressure on weight loss of high-volatility bituminous coal heated in hydrogen and helium atmospheres. Data from Anthony [144].

light hydrocarbon gases such as methane. Quantitative characterization of this model was represented in terms of final yields after complete conversion of unreactive char as well as in terms of the kinetics describing intermediate reaction stages.

The nomenclature in Table 13 defines the different classes of volatiles formed or evolved in the model proposed by Anthony and Howard [146]. Using this notation, a material balance performed on the reactive volatiles within the coal particle leads to the following expression:

$$\frac{dV'_r}{dt} = \frac{k_c}{P}C + k_1 C + k_2 P_{H_2} C + \frac{dC}{dt} \qquad [51]$$

where

C = concentration of reactive volatiles in particle

k_c = overall coefficient for mass transfer of reactive volatiles out of particle

KINETICS OF GASIFICATION REACTIONS 77

k_1 = overall rate constant for deposition of reactive volatiles within particle

k_2 = overall rate constant for stabilization of reactive volatiles within particle, by gaseous hydrogen

$\dfrac{dV_r'}{dt}$ = rate of formation of reactive volatiles

P = total pressure

P_{H_2} = hydrogen partial pressure

Assuming that the derivative dC/dt is approximately equal to 0 and that the ratios of rate constants and transfer coefficients are independent of temperature, the total yield of evolved reactive volatiles, for reaction times approaching infinity, is given by

$$V_r^* = \frac{V_r^{**}[(1/P) + (k_2/k_c)P_{H_2}]}{[(1/P) + (k_2/k_c)P_{H_2} + (k_1/k_c)]} \qquad [52]$$

As indicated by Equation 52, this model assumes that the total yield of evolved reactive volatiles is independent of the time–temperature history, but that it increases with increasing hydrogen partial pressure due to stabilization of reactive volatiles and decreases with increasing total pressure due to increased diffusion resistance. In the model the total weight loss is the sum of the nonreactive volatiles evolved, the reactive volatiles evolved, and the amount of carbon that is directly gasified with hydrogen concurrent with the formation of reactive volatiles. The weight-loss fraction attributed to this latter process is assumed to be expressed by the term $k_3 P_{H_2}$, thus

$$V^* = V_{nr}^{**} + V_r^* + k_3 P_{H_2} \qquad [53]$$

TABLE 13 NOMENCLATURE FOR MODEL OF ANTHONY AND HOWARD [146]

V_r' = Reactive volatiles formed up to time t
V_r^{**} = Reactive volatiles formed up to $t = \infty$ (potential ultimate yield)
V_r^* = Reactive volatiles lost from particles up to time $t = \infty$ (ultimate yield)
V_{nr}^{**} = Nonreactive volatiles formed and lost from particles up to time $t = \infty$ (ultimate yield)
V = Total volatiles lost from particles up to time t
V^* = Total volatiles (reactive and nonreactive) lost from particle up to time $t = \infty$

Analysis of data obtained in pure helium, in pure hydrogen, and in hydrogen–helium mixtures led to the following evaluation (in Equations 52 and 53) of kinetic parameters, which were made to describe final weight losses obtained with bituminous coal at 1000°C, over the total range of conditions employed:

$$V_r^{**} = 0.20 \tag{54}$$

$$V_{nr}^{**} = 0.372 \tag{55}$$

$$\frac{k_2}{k_c} = 0.02 \text{ atm}^{-2} \tag{56}$$

$$\frac{k_1}{k_c} = 0.56 \text{ atm}^{-1} \tag{57}$$

$$k_3 = 0.001 \text{ atm}^{-1} \tag{58}$$

The rate constant, k_3, was not actually determined from the data developed by Anthony, Howard, et al. [145, 146] but was estimated from the high-pressure data of Moseley and Paterson [140]. This assumption has a strong effect on the evaluation of the parameter k_2/k_c and thus strongly affects the relative amounts of evolved carbon that are attributed to (1) hydrogenation of the reactive volatiles within the particle, which is stoichiometrically limited and is likely to lead to heavy hydrocarbon products and (2) direct hydrogenation of the coal, a process likely to lead to light gaseous hydrocarbons [157].

As discussed previously, Zahradnik and Glenn [157] and Zahradnik and Grace [160] used a different form of correlation to describe the methane yields obtained in Moseley and Paterson's study, so that the carbon conversion due to the direct reaction of coal and hydrogen is given by

$$Y_{CH_4} = Y_{CH_4}^0 + \frac{(1 - Y_{CH_4}^0)bP_{H_2}}{1 + bP_{H_2}} \tag{59}$$

where

$$Y_{CH_4}^0 = 0.08 \tag{60}$$

$$b = 0.006 \text{ atm}^{-1} \tag{61}$$

This expression can be converted into the weight-loss fraction for the bituminous coal used by Anthony and colleagues by assuming a reasonable value of feed carbon in the raw coal of 0.7 g/g of coal, yielding the expression

$$\left(\frac{\Delta W}{W_0}\right)_r = \frac{0.00386 P_{H_2}}{1 + 0.006 P_{H_2}} \quad [62]$$

where the left-hand side is the weight fraction of the raw coal lost by direct coal hydrogenation. If it is assumed that Equation 62 represents the contribution of the direct reaction of coal and hydrogen in Anthony's experiments rather than the term $k_3 P_{H_2}$, and that $(k_2/k_c) = 0$ rather than 0.02, as assumed by Anthony and co-workers, then in terms of yields due to coal-hydrogen reaction and yields due to stabilization of reactive volatiles, there is a significantly different interpretation of total weight loss. This is shown in Figure 34, where case I refers to the Anthony group's evaluation of parameters and case II refers to the use of Zahradnik and Glenn's formulation for direct reaction of coal and hydrogen, with $k_2 = 0$. Although the total weight loss of case II is somewhat less than that

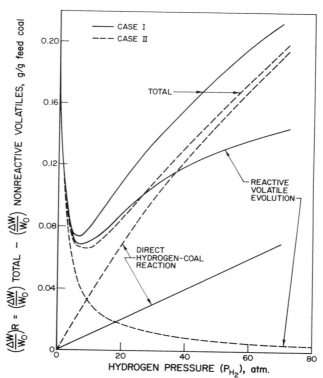

Figure 34 Comparison of reaction distributions.

80 FUNDAMENTALS OF COAL GASIFICATION

obtained by Anthony and colleagues, this could be explained partly by the fact that the evaluation of b by Zahradnik and Glenn [157], using Equation 59, was based on the hydrogen pressure in the feed gas. Since there were substantial average concentrations of methane in Moseley and Patterson's tests, use of average or product-gas hydrogen pressures would result in an increased value of b, which would tend to increase the computed total weight loss for case II (Figure 34). In any event, it is apparent that the quantitative distribution of products evolved during the gasification of hydrogen, interpreted with respect to the model proposed by Anthony and colleagues should be considered tentative, based on the experimental evidence considered. The qualitative features of this model, in terms of the reaction scheme depicted in Figure 27a indicate that tars and oils are evolved by reactions IV and VII, with production of gaseous hydrogen favoring reaction VII over the competing deposition reaction VIII. Light hydrocarbon production is favored by gaseous hydrogen via reaction II.

The rates of formation of nonreactive and reactive volatiles have also been quantitatively described by Anthony and colleagues, assuming first-order kinetics but using a distribution of activation energies for the reacting species. The specific formulation proposed leads to the following expression to define total coal weight loss at any time during the hydrogenation process

$$V = V_{nr}^{*=} \left\{ 1 - C_1 \int_0^{E_1} f(E) \exp - \left[\int_0^t k_0^0 \exp\left(-\frac{E}{RT}\right) dt \right] dE \right\}$$

$$+ (V_r^* + k_3 P_{H_2}) \left\{ 1 - C_2 \int_{E_1}^{\infty} f(E) \exp - \left[\int_0^t k_0^0 \exp\left(-\frac{E}{RT}\right) dt \right] dE \right\}$$

[63]

where

$f(E)$ = activation energy distribution function (Gaussian)

$$= [(2\pi)^{1/2} \sigma]^{-1} \exp \frac{-(E - E_0)^2}{2\sigma^2} \qquad [64]$$

$$C_1 = \left[\int_0^{E_1} f(E) \, dE \right]^{-1} \qquad [65]$$

$$C_2 = \int_{E_1}^{\infty} f(E) \, dE \qquad [66]$$

Equation 65 was fitted to the data of Anthony and co-workers, using the values for limiting yields indicated in the preceding equations and the following values for other parameters:

k_0^0 = 1.67 × 10^{13} sec^{-1}
E_0 (mean activation energy) = 54.8 kcal/g-mol
σ (standard deviation) = 17.2 kcal/g-mol
E_1 (minimum activation energy for formation of reactive volatiles) = 61.4 kcal/g-mol

Results obtained by the Anthony group for the gasification of lignite showed that although weight loss increased with increasing hydrogen partial pressure, weight loss did not decrease with increasing total pressure, as was observed with bituminous coals at pressures up to 10 atm. This implies that, with lignite, either no reactive volatiles capable of repolymerizing on coal surfaces are formed or the diffusion of the reactive volatiles that are formed is extremely rapid, even at elevated pressures. Although both explanations are likely, diffusion resistance is supported by the fact that gaseous diffusion with bituminous coals would be particularly inhibited during the partial liquefaction that occurs at intermediate devolatilization stages, a phenomenon that does not occur with lignites.

The effects of diffusion inhibition on initial coal-gasification behavior have received relatively little attention in studies conducted so far. In one series of experiments by Anthony and colleagues (in hydrogen at 1000°C and 69 atm), total weight loss obtained with bituminous coal for residence times of 5-20 sec was found to increase by 44-59% of the original weight of the coal as the diameter of the original coal particle decreased from 1000 μ to 70 μ. This was interpreted as reflecting an increased internal mass-transfer resistance with increased particle size, which inhibited diffusion of external hydrogen into the coal-particle interstices, thus reducing the effective hydrogen partial pressure. There is a possible complication in the qualitative interpretation of the results of Anthony and co-workers, however, due to the use of a fixed bed of bituminous coal particles, which can agglomerate during devolatilization and significantly increase effective particle sizes. For example, with lignite, a nonagglomerating "coal," the Anthony group found no effect of particle size on weight loss.

Moseley and Patterson [141], using a nonagglomerating coal, also obtained some indication of enhanced gasification yields with decreased particle sizes. However, interpretation of these results was complicated by considerable data scatter and by the fact that, in varying particle size from 60 μ to 200 μ, it was necessary to vary heating rate and coal residence times as well. The possibility that larger particles may not reach reactor temperatures in the short residence times characteristic of transport reactor systems could well be a significant factor affecting Moseley and Patterson's results.

82 FUNDAMENTALS OF COAL GASIFICATION

This effect may also apply to results obtained by Hill, Anderson, et al [161] for the catalyzed gasification of a nonagglomerating coal in hydrogen at 137 atm, in a dilute-phase transport reactor system. In that investigation, conducted at 515°C, total coal conversion to liquids and gases decreased from 91% to 29% on an maf when average particle sizes were increased from 40 μ to 420 μ. Although the use of catalysts may drastically alter the gasification process, it is unlikely that the severe effect of particle size observed by Hill and colleagues is due solely to mass-transfer inhibition. Thus it is uncertain at this time what causes particle size to have an effect during the initial stages of gasification in hydrogen, and future detailed studies using more refined experimental techniques are required to distinguish between possible alternatives.

Johnson recently studied the initial gasification stages of low-rank coals in a dilute-phase transport reactor [153-155]. This study is distinct from an earlier investigation using a thermobalance apparatus, discussed previously. The transport reactor consists of a 200-ft-long helical coiled tube that is $\frac{1}{16}$ in. in diameter. The reactor tube itself is the heating element, with electrodes attached at various points along the length of the tube to permit the establishment of desired temperature profiles and thus to preset time-temperature histories for the gas-solid streams passing through the coil. With this system, the gasification kinetics of lignite and subbituminous coal were studied in hydrogen, helium, and hydrogen-helium mixtures as a function of pressure (18-52 atm), gas composition, temperature (500-850°C), controlled rates of heat up to 80°C/sec, and coal gas-residence times (5-15 sec) when the reactor was operated isothermally.

Typically experimental results of this study, expressed as yields of gasification products and as conversions of individual coal components, are shown in Figures 35 and 36. Based on such results, Johnson described the initial gasification of low-rank coals with a two-stage model similar to the models proposed by Zahradnik and Glenn and by Moseley and Patterson. The first stage involves primary devolatilization (pyrolysis) processes related to thermally activated reactions that lead to evolution of the bulk of the volatile matter. As shown in Figure 35 for the gasification of lignite in hydrogen at a relatively slow heat-up rate, the first-stage devolatilization is initiated below 500°C. Coal oxygen is evolved primarily as carbon dioxide, carbon monoxide, and water, with a small amount observed in the phenolic oils.

Evolution of carbon dioxide occurs below 500°C, apparently resulting from the decomposition of carboxyl functional groups in the coal. Decomposition of other oxygenated functional groups leads to the evolution of both carbon monoxide and water, with evolution of the total available oxygen in the coal being completed at about 700°C. This is shown more clearly in Figure 36. Although carbon dioxide yields begin to decrease above 650°C, this was considered to be due solely to a secondary reaction occurring in the reactor tube, probably catalyzed by the reactor walls. Evolution of C_3^+ gaseous hydrocarbons as well as

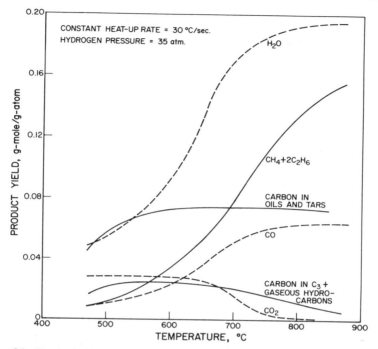

Figure 35 Typical product yields for gasified Montana Lignite in hydrogen. Data from Johnson [153].

formation of oils and tars occur below 550°C, although yields of this species group decrease above 650°C because of the hydrogenation to methane and ethane. Total yields of oils and tars remained relatively constant up to 850°C, although the composition of this fraction changed substantially with increasing temperature, becoming lighter in nature. Conversion of the heavier oils to benzene is initiated at about 650°C, and at 850°C approximately one-half of the total oil-tar fraction consists of benzene.

Johnson considers the first-stage devolatilization reaction to be complete when evolution of the available coal oxygen is completed. For the results shown in Figure 36, this occurs at about 700°C. The solid product of this first-stage reaction is a semichar consisting primarily of carbon and hydrogen. Under pyrolysis conditions in an inert atmosphere, this semichar undergoes second-stage reaction by coal–hydrogen evolution, leading to the formation of relatively unreactive char. In the presence of gaseous hydrogen at elevated pressures, however, coal–hydrogen evolution is accompanied by coal-carbon gasification to form methane and ethane. This behavior was interpreted by postulating that the thermally activated processes that result in coal–hydrogen evolution lead to for-

84 FUNDAMENTALS OF COAL GASIFICATION

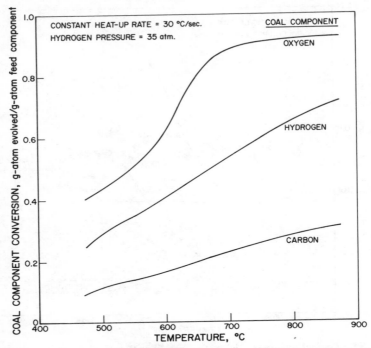

Figure 36 *Typical coal-component conversion during gasification of Montana lignite in hydrogen. Data from Johnson [153].*

mation of an intermediate reactive structure or structure component in the carbon matrix. This structure can either deactivate to form a stable char structure or can react with gaseous hydrogen to form both methane and ethane. When conversion of the semichar to stable char is complete, further carbon gasification occurs at much lower rates characteristic of coal-char-gasification kinetics.

To isolate quantitatively the kinetics of the rapid-rate reactions between gaseous hydrogen and the intermediate coal structure, Johnson found it necessary to estimate the fraction of methane-plus-ethane products formed from other reactions. This was done by assuming that total methane-plus-ethane yields obtained in the gasification of coal results primarily from three overall processes: thermally activated coal decomposition reactions occurring during the initial devolatilization stage, gas-phase hydrogenation of C_3^+ gaseous hydrocarbons evolved during the first stage, and direct rapid-rate coal hydrogenation occurring during the second stage. With a number of approximations, generalized kinetic correlations were proposed to describe the coal decomposition reactions and the gas-phase hydrogenation reactions as functions of residence time, temperature,

KINETICS OF GASIFICATION REACTIONS   85

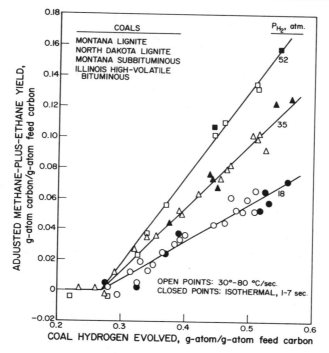

Figure 37   Relationship between adjusted methane-plus-ethane yields and coal-hydrogen evolution for different coals. Data from Johnson [154].

and coal type. With these simplified correlations, total observed methane-plus-ethane yields were adjusted to correspond to first-order estimates of yields due only to direct coal hydrogenation.

Adjusted methane yields correlated strongly with coal-hydrogen evolution for the different coals tested. This behavior is illustrated in Figure 37 [154] for results obtained from the gasification of four different coals with hydrogen. The stoichiometric relationship between coal-hydrogen evolution and adjusted methane-plus-ethane yields depended only on the hydrogen partial pressure and not on the time-temperature history for the different coals. However, for a given coal, the kinetics of coal-hydrogen evolution were dependent only on time-temperature history and not on gas composition or pressure. This is illustrated in Figure 38 [155] for results obtained with Montana lignite. This behavior was formally represented by the following reaction scheme:

$$CH_x^0 \xrightarrow[\text{(thermally activated coal-hydrogen evolution)}]{k_0} CH_y^* \qquad [8]$$

Semichar   Active intermediate

$$\mathrm{CH}_y^* \xrightarrow{k_1} (1-m)\mathrm{CH}_y^0 \text{ Inactive char} \qquad [9]$$

$$\mathrm{CH}_y^* \xrightarrow[\text{(gaseous hydrogen)}]{k_2} m\left[z\mathrm{CH}_4 + \frac{(1-z)}{2}\mathrm{C}_2\mathrm{H}_6\right]$$

where

$$m = \frac{k_2}{k_1 + k_2} \qquad [67]$$

This model is similar to that proposed by Zahradnik and Glenn [157], in which it is assumed that the active intermediate is consumed during hydrogenation. It differs somewhat from Moseley and Patterson's [140, 141] model, which assumes that there is no net consumption of the active intermediate in the hydrogenation step and that the intermediate is lost through thermal deactivation.

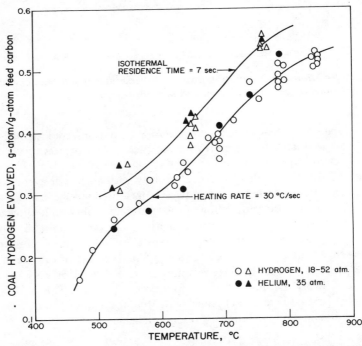

Figure 38 Effect of temperature on coal-hydrogen and in helium. Data from Johnson [155].

Although the distinction as to whether the active intermediate is consumed in the hydrogenation reaction is of theoretical importance in describing the rapid-rate gasification step, this distinction is difficult to make on the basis of existing experimental evidence since similar forms of correlations can be derived using either premise to describe gasification behavior under conventional conditions.

A particularly important feature of Johnson's model compared to others is that it specifically characterizes the active intermediate as a structure resulting from the coal–hydrogen evolution during second-stage devolatilization. This concept was previously proposed by Moseley and Patterson in stating that reactive groupings, which normally yield hydrogen during hydrogasification, are probably the main agents of activation with respect to the rapid-rate methane formation reaction. In Johnson's model, the rate of Reaction 9 was assumed to be very rapid relative to Reaction 8, and the ratio k_2/k_1 was assumed to be independent of temperature. This ratio, however, was found to be directly proportional to the hydrogen partial pressure, apparently reflecting an effect of this partial pressure on the hydrogenation step rather than on the deactivation step. With the assumptions made, the total methane-plus-ethane yield from direct hydrogenation, after completion of deactivation of the active intermediate, is independent of the time–temperature history and depends only on hydrogen partial pressure, according to the following relationship:

$$Y''_{CH_4} = \lambda m \qquad [68]$$

$$m = \frac{k_2/k_1}{1 + (k_2/k_1)} \qquad [69]$$

$$k_2 = k_2^0 P_{H_2} \qquad [70]$$

where

λ = total carbon that forms semichar (g-atom/g-atom carbon in feed coal)

Y''_{CH_4} = carbon in methane-plus-ethane formed in rapid-rate hydrogenation after completion of deactivation of active intermediate (g-atom/g-atom carbon in feed coal)

The form of this expression is very similar to that proposed by Zahradnik and Glenn. Prior to complete deactivation of the active intermediate, the yield of methane-plus-ethane, Y'_{CH_4}, is described by the following more general expression:

$$Y'_{CH_4} = m\lambda f \qquad [71]$$

where f is the fraction of initial semichar that has been reacted by means of Reaction 8. To describe the kinetics of coal–hydrogen evolution that reflect the conversion of semichar to char, Johnson assumed that Reaction 8 was first order and dependent only on the time–temperature history. This overall reaction was considered to involve thermally activated bond-rupturing processes, with a distribution of activation energies to account for the heterogeneous characteristics of the reacting solids. For simplicity, a continuous uniform distribution of activation energies was assumed, leading to the expression

$$1 - f = \int_{E_0}^{E_1} f(E) \left\{ \exp \left[-k_0^0 \int_0^t \exp\left(\frac{-E}{RT}\right) dt \right] \right\} dE \qquad [72]$$

where

$$f(E) = (E_1 - E_0)^{-1}$$
$$k_0 = k_0^0 \exp \frac{-E}{RT}$$

The parameters in Jonson's model were quantitatively evaluated, based on data obtained from the gasification in hydrogen of a Montana lignite, Montana subbituminous coal, and North Dakota lignite, as well as on data reported by Feldman, Mima, et al. [119] for the gasification of Illinois No. 6 high-volatility bituminous coal in methane–hydrogen mixtures. The following evaluations were made to describe results obtained with these coals:

$$k_0^0 = 1.97 \times 10^{10} \text{ sec}^{-1}$$
$$E_0 = 40.8 \text{ kcal/g-mol}$$
$$E_1 = 62.9 \text{ kcal/g-mol}$$
$$\lambda = 0.8 \text{ g-atom/g-atom feed carbon}$$
$$k_2^0/k_1 = 0.0083 \text{ atm}^{-1}$$

These evaluations describe results obtained at coal–gas heat-up rates of 30°C/sec and 80°C/sec, and at isothermal residence times of 1–7 sec.

Johnson also tested the possibile applicability of his model to describe rapid-rate methane formation kinetics with a variety of other bituminous coals, based on data reported in the literature that were obtained in investigations employing dilute-solid-phase reactor systems. These data, obtained over a wide range of conditions, are shown in Figure 39. The values of coal and gas residence times indicated in Figure 39 were reported directly or were deduced from other reported information. Since a significant conversion of feed gases was observed in many of the studies, the hydrogen pressure selected for correlation was the hydrogen partial pressure in the reactor product gases.

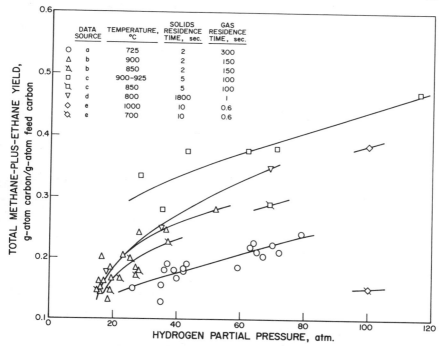

Figure 39 Total methane-plus-ethane yields obtained in different studies on gasification of bituminous coals with hydrogen. Data from Lewis, Friedman, et al. [150], Feldmann, Mima, et al. [119], Moseley and Patterson [141], Hiteshue, Friedman, et al. [143], and Graff, Dobner, et al. [148].

In interpreting the data shown in Figure 39, Johnson computed methane-plus-ethane yields corresponding only to direct coal–hydrogen reactions by subtracting the values estimated by his correlations predicting pyrolysis yields and yields from secondary hydrogenation of C_3^+ gaseous hydrocarbons from total methane-plus-ethane yields. Next it was assumed that the values of Y'_{CH_4} were in agreement with Equation 75, $Y'_{CH_4} = m\lambda f$. Values of m were then computed by dividing Y'_{CH_4} by the values of λ and f predicted by his model, in which $\lambda = 0.8$, and f was computed from Equation 76 for the appropriate temperature and coal residence times reported. The computed values of m, as a function of hydrogen partial pressure, are compared in Figure 40 with Johnson's correlation describing results obtained with subbituminous coals and lignite. In spite of the reasonable consistency shown for hydrogen partial pressures up to 120 atm, Johnson points out that certain of Moseley and Patterson's data at higher pressures, (200–350 atm) show values of m higher than those predicted by the correlation, which suggests that this model may not be applicable in this pressure range.

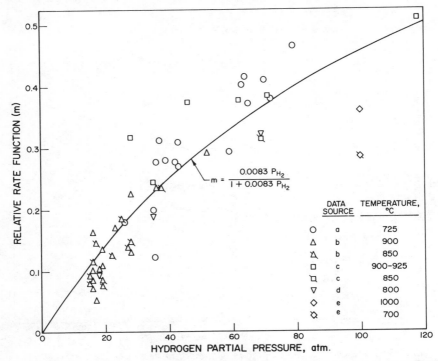

Figure 40 Correlation of methane–ethane yields obtained in various investigations of high-volatility bituminous coal gasification in hydrogen. Data from Lewis, Friedman, et al. [150], Feldmann, Mima, et al. [119], Moseley and Patterson, [141], Hiteshue, Friedman, et al. [143], and Graff, Dobner, et al. [148].

Most of the data shown in Figure 40 were obtained with gases containing substantial concentrations of methane, carbon oxides, and steam, suggesting that these species have at best only secondary effects on the rapid-rate methane reaction. As discussed previously, this is consistent with the fact that correlations proposed by various investigators to describe this reaction have been expressed only in terms of hydrogen partial pressure, independent of the concentration of other individual gaseous species.

An exception to this is the correlation of Anthony, Howard, [145] describing the evolution of reactive volatiles, in which the total carbon gasified also depends on total pressure due to the proposed effect of diffusion inhibition. Nevertheless, it appears that, for practical purposes, the direct coal–hydrogen reaction is irreversible under the ranges of conditions that have been employed in most investigations. This, of course, does not preclude the possibility of secondary reactions (either homogeneous or catalyzed by reactor walls or by the coal itself) that may alter product methane yields, as reported by some investigators [160]. The contribution of secondary reactions under the complex conditions existing in most of the experimental systems used, however, is an area in need of further study.

The assumption of first-order kinetics with a range of activation energies to characterize the controlling step in rapid-rate methane formation appears to be a convenient technique to approximate the complex phenomena that must actually be involved in this reaction. It is important to note, however, that in representing experimental results by a first-order reaction with a range of activation energies, caution should be exercised in interpreting the fundamental processes based on numerical evaluation of the kinetic parameters. For example, in describing the kinetics of coal–hydrogen evolution by a first-order reaction, with a constant activation energy distribution ranging from E_0 to E_1, Johnson has shown the existence of a large covariance between the preexponential factor and the activation-energy parameters [154, 155].

This is illustrated in Figure 41, where variations of the values of conversion, f, calculated using values of the kinetic parameters recommended by Johnson are compared to values of f calculated using a significantly distinct set of kinetic parameters. Nevertheless, very similar characteristics result over a wide range of

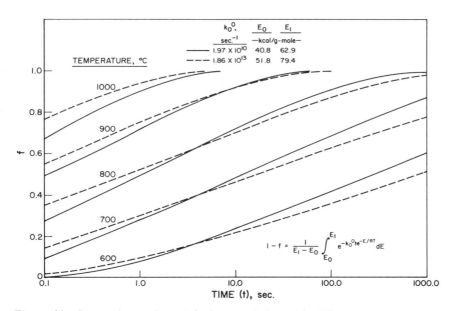

Figure 41 Comparison of model characteristics with different sets of kinetic parameters.

92 FUNDAMENTALS OF COAL GASIFICATION

temperatures and residence times. A much greater range of estimates of kinetic parameters could result when describing data obtained at a single residence time over a wide range of temperatures. It is thus apparent that significantly different numerical evaluations could be made by different investigators assuming an activation energy distribution to characterize the same first-order reaction, even if the data considered are apparently consistent.

As discussed previously, Anthony and colleagues [144-146] proposed that a rate-limiting step in the initial stage of coal hydrogasification is the formation of reactive volatiles—a first-order reaction with a distribution of activation energies—within the coal structure. These investigators represented the activation energy distribution by a function equivalent to the high-energy tail of a Gaussian distribution. In Figure 42 this unusual function is compared to the constant distribution function proposed by Johnson to describe a possibly equivalent reaction step, namely, coal-hydrogen evolution.

Because of differences in the empirical forms used by these investigators, in addition to differences in the activation energy ranges, it is of interest to compare the predictions of the models proposed to describe carbon gasification yields during coal hydrogasification. Anthony's model describes coal weight loss

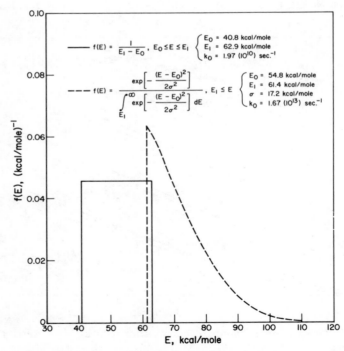

Figure 42 Comparison of activation-energy distribution functions.

due to the evolution of reactive volatiles and direct coal-hydrogen reaction as a function of hydrogen pressure and time-temperature history. If it is assumed that 95% of this weight loss is due to carbon gasification and that the feed coal has a carbon content of 70%, the predicted weight losses can be translated into carbon-conversion fractions, not including the carbon in the nonreactive volatiles gasified. If it is further assumed that this carbon conversion fraction corresponds to rapid-rate methane formation, the predictions of Anthony's model can be directly compared to the predictions of Johnson's model describing methane formation due solely to coal-hydrogen reaction.

This comparison is shown in Figure 43 as a function of residence time for gasification in hydrogen at 20 atm and 50 atm and at 900°C. Although residence times up to 100 sec are shown, the data in both investigations considered were generally obtained at coal residence times less than about 20 sec. Nevertheless, both models predict that relatively little conversion occurs after 20 sec at 900°C. In addition, the magnitude of conversion predicted by the two models is similar, particularly at the lower hydrogen pressure, in spite of the differences in the models and correlations applied. This is also true at other temperatures, as shown in Figure 44, which compares the predictions as a function of temperature at a residence time of 10 sec.

The controlling reaction stages in the models proposed by Anthony and by Johnson are characterized by relatively high activation energies (>40 kcal/g-mol). In other investigations simpler kinetic models have been proposed to describe rapid-rate methane formation, which have led to relatively low activation

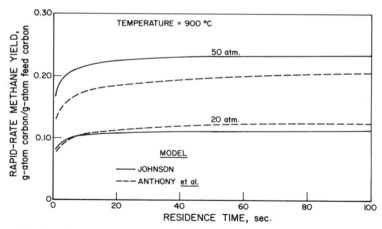

Figure 43 Comparison of predicted methane yields as a function of residence time and hydrogen pressure, based on different models. Data from Johnson [153-155], Anthony [144], Anthony, Howard, et al. [145], and Anthony and Howard [146].

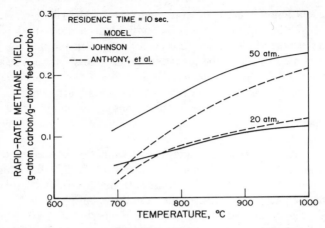

Figure 44 Comparison of predicted methane yields as a function of temperature and hydrogen pressure, based on different kinetic models. Data from Johnson [153-155], Anthony [144], Anthony, Howard, et al. [145], and Anthony and Howard [146].

energies. Feldman [119], for example, proposed a model in which rapid-rate methane formation rates are directly proportional to the hydrogen partial pressure and to the concentration of solid carbon participating in the reaction. This model, like those proposed by Moseley and Patterson and by Zahradnik and Glenn, assumes that at elevated temperatures devolatilization occurs essentially instantaneously, followed by the slower but still rapid reaction of hydrogen and coal to form methane. Feldmann's model does not, however, consider deactivation of the reactant intermediate.

In analyzing data obtained at the U.S. Bureau of Mines using Feldmann's model (characterized by a single first-order rate constant), Gray, Donatelli, et al. [162] have obtained an activation energy of 15.1 kcal/g-mol. A similarly low activation energy of 15 kcal/g-mole was determined by Zahradnik and Glenn for coal hydrogenation in a dilute-phase flow reactor. Although these relatively low activation energies have been interpreted as representing the difference between activation energies for gasification and deactivation of reactive intermediates, or as indicating the effects of mass-transfer limitations, such values are not inconsistent with the much higher activation energies proposed in the models that employ ranges of activation energies. If, for example, values of f predicted from Johnson's model at a single residence time are plotted against reciprocal temperature, the slope of the line can be interpreted as an overall activation energy according to the following expressions:

$$\frac{df}{dt} = k(1 - f) \qquad [73]$$

$$k = k^0 \exp \frac{-E}{RT} \qquad [74]$$

$$-\ln(1-f) = kt = k^0 t \exp \frac{-E}{RT} \qquad [75]$$

$$\ln[-\ln(1-f)] = \ln k^0 t \frac{-E}{RT} \qquad [76]$$

Such Arrhenius plots are shown in Figure 45 for residence times of 1 sec, 10 sec, and 100 sec, resulting in activation energies of 13.7–15.6 kcal/g-mol. This range of values is quite consistent with the apparent activation energies reported by Zahradnik and Glenn and by Gray, based on data obtained at similar coal residence times.

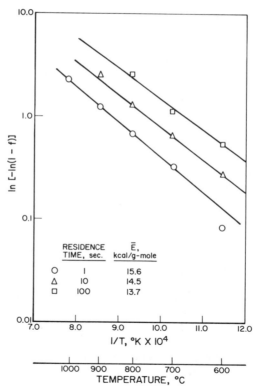

Figure 45 Arrhenius plots for constant-residence-time conversion function.

Coal-char-gasification Kinetics

Although the initial coal-gasification stage is completed in seconds or less at elevated temperatures, the subsequent gasification of the coal chars produced during this initial reaction stage is much slower, requiring minutes or hours to obtain significant conversions under practical conditions. Since reactor design criteria and volume requirements for commercial gasifiers are largely dependent on coal-char, reactivity, the kinetics of coal-char-gasification systems have been extensively investigated.

One advantage of studying coal-char-gasification kinetics, compared to initial coal reaction kinetics, is that it is relatively easy to conduct experiments under conditions of differential gas conversion, whereby the composition of gas in contact with the coal char is known and remains constant. This greatly facilitates kinetic analysis of the experimental results. This procedure has usually been applied to fixed-bed flowing gas systems, which have been used in the majority of experimental studies to obtain detailed kinetic information about gasification kinetics of coal chars.

A few studies have also begun to use fluidized beds to obtain differential gasification rates, although this technique has been criticized because of the potential heterogeneous characteristics of such a system [3]. The fluidized bed, however, does have the advantage of good circulation of solids, leading to uniform bed temperatures. With fixed beds, temeprature control is a greater problem because of the strong heat effects that can occur during coal-char gasification, an effect that can lead, when undefined, to apparently inconsistent results obtained in different experimental studies. Dilute solid-phase transport systems have not been employed to study coal-char reactions because of the negligible carbon conversions that would occur during the short residence times for solids characteristic of such systems.

Other complications that must be considered in interpreting data obtained from coal-char-gasification studies are related not only to the different coals used but also to differences in the procedures employed in preparing chars from these coals. For example, the term "coal char" is usually defined as the residue resulting from any heat treatment of coal. If, however, the pretreatment temperature and residence times are sufficiently low, the resultant coal char can exhibit some rapid-rate methane formation reactivity during subsequent gasification. In the following discussions the term "coal char" is intended to refer primarily to residues that have been subjected to preparation conditions sufficiently severe that negligible rapid-rate activity remains. As is discussed later, even with this limitation, pretreatment conditions can still affect the subsequent reactivity of coal chars during gasification.

The following coal-char-gasification reactions have received the greatest attention in recent years because of the importance of developing commercial processes to convert coal to high- or low-energy gases:

$$C + 2H_2 \Longleftrightarrow CH_4 \qquad [1]$$

$$C + CO_2 \Longleftrightarrow 2CO \qquad [2]$$

$$C + H_2O \Longleftrightarrow CO + H_2 \qquad [10]$$

The carbon–hydrogen reaction (Reaction 10) has been studied almost solely at elevated pressures because of the strongly favorable effect of pressure on this reaction and the almost negligible rates obtained at atmospheric pressure. This reaction is particularly important to the production of high-energy gas, whereas Reactions 2 and 10 are important to the production of both high- and low-energy gases. Although Reactions 1 and 2 can each occur alone in gasification systems, roughly according to the stoichiometry indicated, the steam–carbon reaction (Reaction 10) cannot generally occur alone under conditions of integral gas conversion. Because of the water-gas shift reaction (Reaction 3)

$$H_2O + CO \Longleftrightarrow CO_2 + H_2 \qquad [3]$$

carbon dioxide can be formed even with a pure steam feed gas and can react competitively with the steam–carbon reaction according to Reaction 2 at low or high pressures. Since hydrogen is a product of the steam–carbon reaction, as well as the water-gas shift reaction, methane can also be formed in steam–carbon systems, at sufficiently high pressures, by a reaction such as Reaction 1.

The representations given by Reactions 1, 2, and 10 are oversimplified for the purpose of illustration since the hydrogen present in coal chars will alter reaction stoichiometry in real systems. Furthermore, the reactions only describe overall stoichiometry in limiting cases and do not imply fundamental reaction steps. In gases containing both steam and hydrogen, for example, methane could conceivably be formed by the direct interaction of carbon, steam, and hydrogen by means of a reaction such as

$$2C + H_2 + H_2O \Longleftrightarrow CO + CH_4 \qquad [11]$$

A variety of other reactions can also be postulated to occur in coal-char-gasification systems, which, in the absence of molecular oxygen, consist of some stoichiometric combination of Reactions 1, 2, and 10. In complex gasification systems containing steam, hydrogen, and carbon oxides, however, most experiments have determined only overall reaction stoichiometry. Although individual reactions have been variously postulated to rationalize this type of result to aid in kinetic formulations, such postulated reaction schemes are likely to be more arbitrary than those that could be obtained with more sophisticated techniques, such as the use of isotopic tracers or reactions with atomic gaseous species.

Gasification reactions have usually been studied under isothermal conditions, often with differential gas conversion. To a lesser extent, experiments have also been made under conditions of constant heat-up rate in order to obtain information on temperature effects in a single experiment. In most studies, the carbon gasification rate, W, has been defined by the expression

$$W = \frac{dX/dt}{1 - X} \qquad [77]$$

where

X = carbon conversion fraction initially present (in mol/mol carbon initially present)

t = time

Using this expression, in which W is sometimes referred to as the "specific" carbon gasification rate, implies that if absolute carbon gasification rates (moles of carbon gasified per unit time) under constant environmental conditions are proportional to the amount of carbon present at any time, W will remain constant with increasing conversion, X. This has frequently been found true, at least for carbon conversion fractions up to about 80% (although in a number of studies some transient behavior has been observed at levels of carbon conversion less than about 10%).

At carbon conversions greater than 80%, the limited experimental information available suggests that specific gasification rates may decrease somewhat. The approximation of relatively constant specific gasification rates over a large range of carbon conversions is at best applicable only to the gasification of chars in the absence of significant catalytic effects from natural or added impurities. The activity of catalysts present can vary substantially during the course of gasification, even under constant environmental conditions, leading to significant variations in specific gasification rates. Devolatilization of fixed beds of bituminous coals can also subsequently lead to variations in the specific gasification rates of chars formed, due to the agglomerating properties of this type of coal. Agglomeration of coal particles can occur during devolatilization, leading to the possible inhibition of gaseous diffusion during the subsequent gasification. When this occurs, initial specific gasification rates are relatively low, but they increase with increasing conversion as restrictions are removed and the internal char surface becomes more available to the reacting atmosphere.

The following discussion of experimental work on char gasification focuses on three main reaction systems: (1) the char–hydrogen–methane system, which is relevant Reaction 10, (2) the char–carbon dioxide–carbon monoxide system, relevant to Reaction 2, and (3) the char–hydrogen–steam–carbon monoxide–carbon dioxide–methane system, relevant to all of the gasification reactions considered, including the steam-carbon reaction (Reaction 10). Although this discussion focuses primarily on the behavior of coal chars, pertinent results obtained with other carbon solids are also considered.

Coal-char-Hydrogen-Methane System. For the gasification of coal chars and other carbons in pure hydrogen, specific carbon gasification rates have been correlated with two main forms of equations:

$$W = \frac{k_H P_{H_2}^2}{1 + K_H P_{H_2}} \quad [78]$$

[7, 137, 159, 163, 164] or

$$W = k_H^* P_{H_2} \quad [79]$$

[137, 164-169]

where

k_H, K_H, k_H^* = kinetic parameters that depend on temperature and the nature of the char, often expressed by an Arrhenius form of expression

P_{H_2} = hydrogen partial pressure

At sufficiently high hydrogen partial pressures, Equation 78 reduces to Equation 79, with:

$$k_H^* = \frac{k_H}{K_H} \quad [80]$$

Based on either of these correlation forms, gasification rates increase significantly with increasing hydrogen partial pressure.

Blackwood and co-workers have proposed a mechanism to rationalize these correlation forms, based on results of several fixed-bed studies conducted during 1959-1967, using coconut char, wood char, brown coal, and Australian bituminous coal chars [137, 164-167]. The following fundamental reaction steps are assumed:

Step 1. $\quad C_f + H_2 \underset{k_1'}{\overset{k_1}{\rightleftarrows}} C(H_2) \quad$ [12]

Step 2. $\quad C(H_2) + H_2 \underset{k_2'}{\overset{k_2}{\rightleftarrows}} CH_4 + C_f \quad$ [13]

where

C_f = active center on carbon surface
$C(H_2)$ = surface complex formed by the chemisorption of hydrogen or by decomposition of methane at an active center

According to this mechanism, gaseous hydrogen chemisorbed at an active center on the carbon surface can react with hydrogen to form methane and to regenerate an active center. The individual reaction steps are reversible. Under steady-state conditions, the following rate equation can be derived based on steps 1 and 2 (Reactions 12 and 13):

$$W = \frac{dX/dt}{1-X} = \frac{n_T(k_1 k_2/k_1') P_{H_2}^2 [1 - (k_1' k_2'/k_1 k_2) P_{CH_4}/P_{H_2}^2]}{1 + (k_1 + k_2/k_1') P_{H_2} + (k_2'/k_1') P_{CH_4}} \quad [81]$$

where n_T is the total number of active centers per mass of carbon.

Blackwood and McCarthy suggest that variations in the total number of active centers, n_T, contribute primarily to differences in the reactivity of different coal chars, implying that the relative behavior of different coal chars should be similar [137]. He has also considered a different set of reaction steps in place of step 1 (Reaction 12) to describe the hydrogen chemisorption process [170]:

Step 3. $\qquad C_f + H_2 \underset{k_3'}{\overset{k_3}{\rightleftarrows}} (CH)(H) \qquad [14]$

Step 4. $\qquad (CH)(H) \underset{k_4'}{\overset{k_4}{\rightleftarrows}} C(H)_2 \qquad [15]$

Steps 3 and 4 (Reactions 14 and 15) are based on the postulate that the main function of active sites on the carbon surface is to split the hydrogen molecule, so that one hydrogen atom becomes attached to the active center and the other hydrogen atom reacts either with the original active site or with another active site to form a $-CH_2$ group. The formation of hydrogen atoms in the chemisorption of hydrogen on carbon surfaces was proposed by Barrer [171] in 1935, who suggested that these atoms are chemisorbed on adjacent dual sites. It was further shown that the chemisorption and desorption of hydrogen at a carbon surface were first-order processes that could be expressed by a Langmuir type of adsorption isotherm. However, Blackwood argues that if hydrogen is adsorbed on dual sites, the desorption would be a second-order process, contrary to experimental results, unless the fraction of surface covered by the two hydrogen atoms could be regarded as fixed—no longer available for chemisorption but able to be reoccupied by the breakdown of a hydrogen molecule.

This requirement for the formation of surface hydrogen atoms as a prerequisite to the formation of methane is supported, although not necessarily proved, by experiments in which various carbons have been reacted directly with hydrogen atoms produced in microwave or radiofrequency fields. Such studies, usually conducted at pressures less than 1 mm of mercury and at temperatures below 300°C, have shown gasification rates equivalent to those obtained in molecular hydrogen at elevated pressures and at temperatures several hundred degrees higher.

Although Blackwood had earlier favored the mechanism given by steps 3 and 4 (Reactions 14 and 15) as an explanation for the formation of -CH_2 groups on the carbon surface [170], he later suggested that step 1 (Reaction 12) was a more likely path [137] and also proposed alternative explanations to characterize the nature of active sites responsible for the chemisorption step. In earlier studies, he found a strong correlation between gasification rate and the concentration of residual oxygen in the carbon [167]. He suggested that this oxygen, combined as chromene or benzpyran structures, was responsible for activating certain sites in the carbon crystallite. Blackwood and McCarthy later suggested that some atoms at the edges of or at defects in the carbon structure may be chemically unsaturated, able to act as centers for initiating molecular dissociation or redox reactions, and that they represent free or unpaired electrons [137].

Zielke and Gorin [163] have proposed a somewhat different reaction mechanism, which they used to develop a kinetic correlation in the form of Equation 78 to describe gasification of a Disco char, derived from commercial processing of a Pittsburgh seam bituminous coal. This mechanism is based on the hydrogenation of exposed edges in the carbon lattice that contain hydrogen atoms. The first step proposed is

$$H_2 + \begin{array}{c} H \\ \diagdown \\ C=C \\ \diagup \quad \diagdown \\ \quad H \end{array} \underset{k'_s}{\overset{k_s}{\rightleftarrows}} \begin{array}{c} H \quad H \quad H \quad H \\ \diagdown \diagup \quad \diagdown \diagup \\ C \text{---} C \\ \diagup \quad \diagdown \end{array} \qquad [16]$$

which is similar to the reaction given by step 1 (Reaction 12). Subsequent reactions in the model involve the progressive hydrogenation of the -CH_2 groups to -CH_3, and then to CH_4, with the regeneration of a new

$$\begin{array}{c} H \quad \quad H \\ \diagdown \quad \diagup \\ C=C \\ \diagup \quad \diagdown \end{array} \qquad [17]$$

grouping for each one consumed. In deriving Equation 78, it was assumed that the hydrogenation of the -CH_3 group, which is effectively equivalent to the reaction given by step 2 (Reaction 13) is rapid compared to other reaction steps.

Based on existing evidence, Equation 81 appears to be the best general kinetic representation of experimental results currently available for steady-state gasification conditions where specific gasification rates are relatively constant during the course of conversion. Some simplification of this expression, however, has been assumed in various applications. The $(k_2'/k_1')P_{CH_4}$ term in the denominator is usually either ignored or assumed to be negligible compared to other terms. In addition, the $(k_1 + k_2)/k_1'$ term in the denominator has been found to be suitably described by a single Arrhenius expression to account for temperature effects. Blackwood and McCarthy have suggested that this probably indicates that the rate constant k_2 is significantly larger than k_1 under most conditions [116]. They further suggest that under most practical gasification conditions, the term $k_2 P_{H_2}/k_1'$ is large compared to unity, and that the rate expression reduces to

$$W = \frac{dX/dt}{(1-X)} = n_T(k_1)P_{H_2}\left[1 - \left(\frac{k_1' k_2'}{k_1 k_2}\right)\frac{P_{CH_4}}{P_{H_2}^2}\right] \quad [82]$$

This, in turn, reduces to Equation 79 for the gasification in pure hydrogen, where

$$k_H^* = n_T k_1 \quad [83]$$

Other investigators, however, have found reaction orders between one and two with respect to hydrogen pressure, at pressures greater than 1 atm, that were best described by Equation 78, at least for pressures less than 20 atm and temperatures up to 900°C [7, 159, 163, 164]. Johnson [7] correlated hydrogen gasification data for air-pretreated bituminous coal char with an expression of the form given by Equation 78, with the following evaluations of the parameters k_H and K_H as a function of temperature:

$$k_H = \exp\left[2.6741 - \frac{27170}{RT}\right] \quad [84]$$

$$K_H = \exp\left[-10.452 + \frac{22050}{RT}\right] \quad [85]$$

where k_H is in the units min^{-1} atm^{-2}, K_H is in atm^{-1}, R is in cal-mol-°K^{-1}, and T is in °K.

The preceding evaluation of k_H is restricted to bituminous coal chars prepared at the same temperature as the subsequent gasification in hydrogen and is based on results obtained at temperatures of 800–1050°C with hydrogen pressures up to 70 atm.

In Figure 46 the effect of hydrogen pressure on specific gasification rates, as predicted by Johnson's [7] correlation at 927°C, is compared to rates ob-

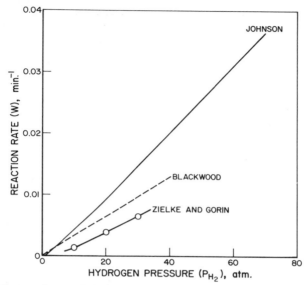

Figure 46 Comparison of hydrogen pressure effects on specific gasification rates. Data from Johnson [7], Blackwood and McCarthy [137], and Zielke and Gorin [163].

tained by Zielke and Gorin [163] for the gasification of Disco char and to rates interpreted from Blackwood's data for the gasification of brown-coal char [137]. In representing Blackwood's results, obtained at 750°C, 850°C, and 950°C and 25 atm pressure, it was assumed that the gasification rates are directly proportional to hydrogen pressure (Equation 79) and that the activation energy for the rate constant k_H^* is about 30 kcal/mol, as suggested by these investigators. Although the different curves represent different reactivity levels for the coal chars used, it can be seen that the nonlinearity exhibited by Johnson's correlation and Zielke and Gorin's data is relatively unimportant, except at low pressure, and that a reasonable approximation of the reactivities of different coal chars can be made by assuming the form of correlation given in Equation 79.

Figure 47 presents a more general comparison of rate constants k_H^* determined from data obtained or correlations proposed in several investigations of the char-hydrogen reaction [7, 137, 163, 165-167, 169, 172, 173]. Pertinent information relevant to these investigations is given in Table 14. The range of values of k_H^* in Figure 47, from Johnson's correlation at a given temperature, is due to the fact that his correlation is of the form given in Equation 78 and us

$$k_H^* = \frac{k_H P_{H_2}}{1 + K_H P_{H_2}} \qquad [86]$$

104 FUNDAMENTALS OF COAL GASIFICATION

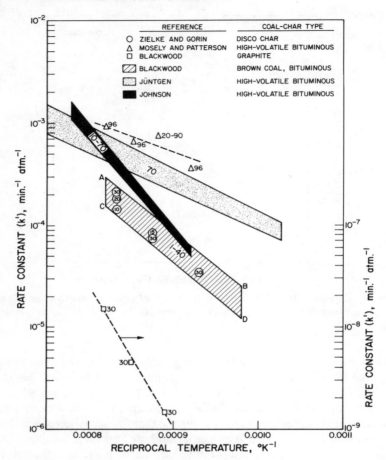

Figure 47 Comparison of rate constants for char gasification in hydrogen. Data from Zielke and Gorin [163], Moseley and Patterson [140], Blackwood [167, 172], Blackwood and McCarthy [137], Blackwood, McCarthy, et al. [165], Jüntgen, Van Heek, et al. [173], and Johnson [7].

This range of values tends to decrease with decreasing temperature since K_H increases with decreasing temperature.

The range of values indicated for Blackwood's results reflects the use of different carbon solids, including wood char, coconut char, brown-coal char, and bituminous-coal chars. Rate constants obtained with these different materials vary by a factor of approximately two at a given temperature and are generally less than values predicted by Johnson's correlation at temperatures greater

TABLE 14 CONDITIONS FOR INVESTIGATIONS OF CHAR–HYDROGEN REACTION

Reference	Coal Type	Primary Particle Diameter Range (μ)	Temperature Range (°C)	Pressure Range (atm)	Maximum Carbon Conversion Fraction	Char-preparation Conditions
169	High-volatility bituminous	150–300	815–950	20–100	~0.3	700°C N_2
173	High-volatility bituminous	—	Ambient–1100°C[a]	70	~0.9	—[a]
7	High-volatility bituminous	420–840	815–1040	10–70	0.9	Reaction temp. N_2
163	Disco char	150–230	815–930	10–30	0.8	Commercial
172	Graphite (reactor grade)	1100–2200	850–950	30	—	Commercial
137	Brown coal	300–760	750–950	25	~0.5	750°C N_2
166	Brown coal	300–760	750–950	25	—	Reaction temp. N_2
165	Bituminous	1100–2200	750–950	20	—	Reaction temp. N_2
167	Coconut char	1100–2200	650–870	2–40	0.1	950°C N_2

[a]Coal heated at constant rate of 10°C/min in hydrogen.

than about 850°C. This difference may be partly due to the effects of coal rank on the reactivity of coal chars for gasification with hydrogen, as shown for an extreme case in Figure 47 for results obtained with graphite. Values of k_H^* given in Figure 47 from Jüntgen, Van Heek, et al. [173] were based on correlations derived for the gasification in hydrogen of a German high-volatility bituminous coal in hydrogen exposed to a constant heat-up rate (10°C/min). The range of values shown corresponds to two different consecutive reaction stages of char gasification postulated by these authors, with a different rate constant applicable to each stage.

Moseley and Patterson's [169] values of k_H^* were determined for the gasification of a British noncaking high-volatility bituminous coal char in a unique experimental system. In this system a horizontal bed of coal char, through which hydrogen was continually flowing, was repeatedly and rapidly heated to reaction temperature (~15°C/sec), maintained at a constant temperature for 15 sec, and then rapidly cooled by water quenching of the reactor tube. Reaction-rate determinations were made assuming a 15-sec residence time at temperature. The values of k_H^* obtained with this procedure are higher than any other values shown in Figure 47. A possible explanation for the apparently high values of k_H^* obtained by Moseley and Patterson as well as those reported by Jüntgen and colleagues for temperatures below 950°C are discussed later.

The effect of methane partial pressure on the gasification kinetics of coal chars in hydrogen–methane mixtures is important in any real system. Neglecting the denominator term containing methane partial pressure in Equation 81, the effect of methane on gasification rates appears in a term that represents thermodynamic reversibility. In this sense, the term $K_{1,2} = (k_1' k_2'/k_1 k_2)$ can be considered an equilibrium constant for the coal-char–hydrogen–methane system. Significantly, and perhaps fortuitously, different investigators, including Zielke and Gorin [163], Johnson [7], and Blackwood and McCarthy [137] have successfully correlated experimental results obtained with coal chars by evaluating $K_{1,2}$ as the equilibrium constant for the graphite–hydrogen–methane system. Consequently, Blackwood suggested that this equilibrium constant is not affected by char type. Although others have reported results obtained from the gasification of coal chars in gases containing steam as well as hydrogen that indicate methane concentrations in excess of values computed from the graphite–hydrogen–methane equilibrium constant [3], such results do not necessarily indicate excess carbon-free energies, since competitive reactions involving steam can lead to methane formation. Also, with integral systems that use coal rather than coal char as the feed solids, methane can result from pyrolysis as well as rapid-rate methane formation, both of which are irreversible processes.

Effects of Coal-Char-Preparation Temperatures. There are two major causes of variations in the reactivity of different coal chars during gasification with hydrogen. One relates to the inherent properties of the coals from which the chars were derived and the other, to the conditions employed in preparing the chars, particularly temperature. It has usually been found that char reactivity decreases with increasing preparation temperature, relative to the temperature of subsequent gasification in hydrogen.

Blackwood, McCarthy, et al. [165] have conducted a systematic study to characterize quantitatively the effect of preparation temperature on the reactivity of brown-coal chars gasified in hydrogen at 25 atm. Results of this study are shown in Figure 48, which plots values of the rate constant k_H^* (defined in Equation 79) as a function of preparation and gasification temperatures. In no case shown in Figure 48, in the range of gasification temperatures of 700-950°C and char-preparation temperatures of 750-950°C, was the char gasified at a temperature higher than its temperature of preparation. Values of k_H^* shown in Figure 48 increase systematically with decreasing preparation temperature, at constant gasification temperature, and increase with increasing gasification temperature at constant pretreatment temperatures.

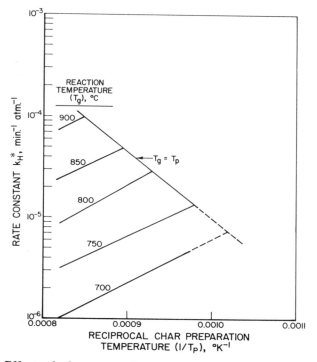

Figure 48 Effect of char-preparation temperature on rate constant for gasification in hydrogen. Data from Gray, Donatelli, et al. [162].

Blackwood and colleagues correlated these results with the following expression:

$$k_H^* = k_{HO}^* \exp\left[\frac{E_p}{RT_p}\right] \cdot \exp\left[-\frac{E_g}{RT_g}\right] \qquad [87]$$

where

T_g = gasification temperature
T_p = preparation temperature $(T_p \geqslant T_g)$
k_{HO}^*, E_p, E_g = constants

For preparation temperatures equal to or less than the gasification temperatures, the value of T_p used in Equation 87 is equal to the gasification temperature, T_g. In rationalizing these results, Blackwood and colleagues proposed that the term $\exp(E_p/T_p)$ reflects the effect of increasing temperature in decreasing the reactivity by the loss of free electrons due to condensation of the carbon structure into more stable condensed ring systems and that previously unstabilized bonds are stabilized in the process. This may be brought about partly through the loss of char hydrogen, which exposes free valences that act as electron traps. This process is thought to be irreversible and to be solely a function of temperature, independent of time after a short period (possibly on the order of seconds). A value of E_p = 20 kcal/mol was determined by Blackwood and co-workers to describe the results given in Figure 48.

For gasification under conditions where $T_g = T_p$, Blackwood and co-workers previously determined a total activation energy for k_H^* of 30 kcal/mol for different coal chars. From Equation 87, then

$$E_g - E_p = 30 \text{ kcal/mol} \qquad [88]$$

and

$$E_p = 20 \text{ kcal/mol} \qquad [89]$$

so that

$$E_g = 50 \text{ kcal/mol} \qquad [90]$$

Based on Equation 82, the value of E_g corresponds to the activation energy of the rate constant k_1; whereas the value of E_p reflects the temperature coefficient pertinent to the concentration of active sites defined by the term n_T.

Johnson [7] has also examined the effect of pretreatment temperatures of

927–1100°C on the rates of gasification of high-volatility bituminous coal char in hydrogen at 927°C and 35 atm. Johnson interpreted his results in terms of Equation 78 and attributed pretreatment effects to decreases in the rate constant k_H. Assuming that Equation 81 is applicable, k can generally be expressed by

$$k_H = k_1 n_T \frac{k_2}{k_1} \qquad [91]$$

Johnson assumed that the variations in k_H with pretreatment temperatures were due to variations in n_T and employing the same correlation form used by Blackwood (Equation 87), determined a value of E_p = 9.340 kcal/mol, about one-half the value determined by Blackwood for brown-coal char. According to Johnson's correlation, the overall activation energy of the term $k_1 n_T (k_2/k_1)$, where $T_p = T_g$, is 27.17 kcal/mol, and the net activation energy for the (k_2/k_1) term is 22.05 kcal/mol. Thus the net activation energy for the term $k_1 n_T$ is 27.17 + 22.05 = 49.22 kcal/mol = $E_g - E_p$. For a value of E_p = 9.34 kcal/mol, then E_g = 58.6 kcal/mol, compared to the value of E_g = 50 kcal/mol, which was based on Blackwood's results with brown-coal char.

The differences in the two values of E_p determined by Blackwood, Cullis, et al. [166] and by Johnson [7] may be due to the different coal chars employed or to the different ranges of char pretreatment temperatures used (Blackwood, 750–950°C; Johnson, 927–1100°C). If the difference is due to differences in the pretreatment temperature, then the two sets of results can be made consistent by using a different empirical form of expression to describe the relative effects of pretreatment temperature. For example, if the term $A \exp(E_p/T_p)$ is normalized to a value of unity at a temperature of 950°C by determining the appropriate values of A for both values of E_p, then a continuous temperature function can be generated, as shown in Figure 49. Obviously, the normalized temperature function cannot be described fully with a term of the form $A \exp(E_p/T_p)$, using a single value of E_p over the total range of temperatures considered, but could be described by a variety of other empirical temperature functions. The curve shown in Figure 49 was computed using the function $(T - T_0)^{-1}$, normalized to a value of unity at 950°C. Although other, more theoretically meaningful forms (which probably require additional parameters) could also be applied, it is apparent that additional experimental information would be useful.

The total concentration of active sites in coal chars is a function of both the total char surface area and the concentration of active sites per unit surface area, as expressed by the relationship

$$n_T = S \cdot \lambda \qquad [92]$$

110 FUNDAMENTALS OF COAL GASIFICATION

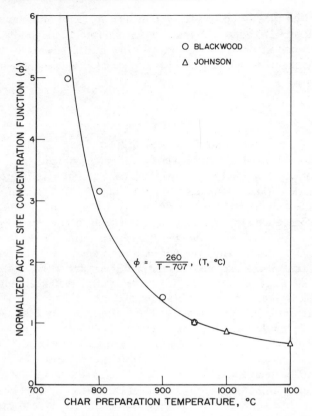

Figure 49 Correlation of active site function with char-preparation temperature. Data from Blackwood, Cullis, et al. [166], and Johnson [7].

where

S = total internal char surface area in m²/mol carbon

λ = concentration of active sites in mol/m²

Thus a decrease in n_T with increasing pretreatment temperature can potentially result from decreasing S and/or decreasing λ. However, unpublished results obtained at the IGT, given in Table 15, show no decrease in the surface area (measured in carbon dioxide) of a lignite char pretreated in nitrogen at temperatures of 927-1100°C and gasified in hydrogen at 927°C. This is consistent with the fact that true graphitization of coal chars does not occur at these temperature levels. Thus the results in Table 15 suggest that the decrease in the reactivity of chars formed at higher temperatures is due to a decrease in the concentration of active sites.

TABLE 15 EFFECT OF DEVOLATILIZATION TEMPERATURE ON INTERNAL SURFACE AREA OF MONTANA LIGNITE CHAR SUBSEQUENTLY GASIFIED IN HYDROGEN AT 927°C[a]

Devolatilization Temperature (°C)	Base-carbon Conversion Fraction of Char	Char Surface Area (m²)	
		g Residue	g Residue (ash free)
927	0.53	740	907
983	0.48	757	905
1038	0.40	774	914
1094	0.30	792	915

[a]Devolatilization in nitrogen, 1 atm, 60 min; char gasification in hydrogen, 35 atm, 30 min.

Johnson [121] has also observed that the internal surface areas do not change during the major course of coal-char conversion when specific gasification rates remain essentially constant, under constant environmental conditions. This implies that both λ and S remain constant with most coal chars, at least up to carbon conversions of approximately 80-90%. One subbituminous coal char tested by Johnson, however, did exhibit continuously decreasing specific gasification rates with increasing carbon conversion; in this case total surface area was found to decrease almost in direct proportion to these rates. Although no reason for these decreasing rates and surface areas was found, it is possible that this particular coal char contained more than one type of carbon, each with a significantly different internal specific surface area. If values of λ were the same for the different carbons, then the carbon types having larger surface areas would react preferentially, resulting in decreasing total gasification rates and total internal surface areas with increasing carbon conversion.

From the available evidence it is probable that decreases in char reactivity resulting from increased pretreatment temperature are due to a decrease in the concentration of active sites, which can be expressed over limited temperature ranges, by the expression

$$\lambda = \lambda_0 \exp(-E_p/T_p) \qquad [93]$$

where λ_0 is a constant that depends on the nature of the particular coal used in char preparation. If Equation 93 applies, the rate of char gasification in hydrogen can be expressed as

$$W = \frac{dX/dt}{1-X} = \frac{k_1^0(S\lambda_0) \cdot \exp(+E_p/RT_p) \cdot \exp(-E_g/RT_g) P_{H_2}^2}{P_{H_2} + k_1'/k_2} \qquad [94]$$

TABLE 16 COMPARISON OF KINETIC PARAMETERS

Correlation (Ref.)	$k_1^0(S\lambda_0)$, atm^{-1} min^{-1}	Activation Energy (kcal/g-mol)		k_1/k_2 (atm)
		E_p	E_g	
7	5.02×10^5	9.4	58.6	$\exp\left[10.452 - \dfrac{22{,}050}{RT}\right]$
137	37.0	20.0	50.0	—

Equation 94 considers the effect of pretreatment temperature, reaction temperature, and hydrogen partial pressure. With respect to the previous discussion, the term $S\lambda_0$ is a property of the feed coal used. A comparison of the parameters in Equation 94 as determined from Johnson's results with coal char from high-volatility bituminous coal [7] and Blackwood and McCarthy's [137] results with coal char derived from brown coal is summarized in Table 16. The ratio of values of the rate constant, k_H^*, predicted by Blackwood's correlation relative to values predicted by Johnson's correlation in the overlapping range of gasification temperatures shown in Figure 47 (line CD for Blackwood) ranges approximately 0.3-0.6, increasing with decreasing temperature. In another study reported by Blackwood [166] in which higher values of k_H^* were obtained with brown-coal char (line AB in Figure 47), the ratio ranges about 0.6-1.1.

Inherent Reactivity of Different Coal Chars. Johnson has determined the gasification reactivity in hydrogen of a large number of coal chars derived from different coal and coal maceral concentrates, varying in rank from anthracite to lignite [121]. Specific gasification rates were obtained at 927°C and 35 atm for coal chars 400-800 μ in diameter, prepared in nitrogen at 927°C (i.e., $T_p = T_g$). Results were expressed in terms of a relative reactivity factor, f_L, defined as

$$f_L = \frac{W}{W_s} \qquad [95]$$

W_s represents the specific gasification rate obtained with a reference coal char. Under the gasification conditions considered ($W_s = 0.0174$ min^{-1}), the values of f_L obtained for all coal chars tested, with the exception of those derived from untreated lignite, were correlated as a function of carbon content, Y, in the raw coals, as shown in Figure 50 [121]. The curve drawn in Figure 50 was computed from the expression

$$f_L = 6.2Y(1 - Y) \qquad [96]$$

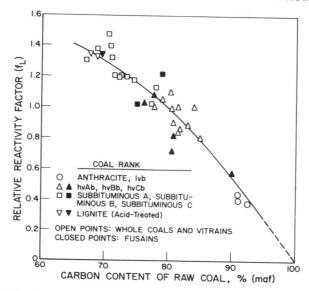

Figure 50 Correlation of reactivity factors. Data from Johnson [121].

for $0.67 \leq Y \leq 0.93$. A similar study was reported by Tomita, Mahajan, et al. [168], who prepared coal chars (40 × 100 mesh) in helium at 1000°C and then determined specific gasification rates in hydrogen at 900°C and 27 atm. Relative reactivity factors (Equation 96) for these gasification rates can be evaluated by defining W_s on the basis of the kinetic parameters given in Table 16 for Johnson's correlation. The values of f_L computed from data obtained with chars derived from anthracite to subbituminous coals are compared to values calculated using Equation 96 in Figure 51 [121, 168]. As indicated, Tomita's data are quite scattered and show only an approximate trend toward the curve of Equation 96.

Transient effects. Transient effects have been observed during the early stages of coal-char gasification in hydrogen in which the specific gasification rates either decrease or increase to a steady-state value [7, 159, 167, 174]. Agglomeration in the char bed during devolatilization of the bituminous coal can be one cause of the initial increase in gasification rates of the coal char. As discussed previously, this can result in a significant increase in the effective particle size, leading to mass-transfer inhibition. Initial gasification can open up additional paths for diffusion, resulting in increased rates, until the char surface becomes completely accessible.

Initially increasing rates can also result from catalytic effects. For example, Walker, Austin, et al. [175] have pointed out that if a carbon initially contains

114 FUNDAMENTALS OF COAL GASIFICATION

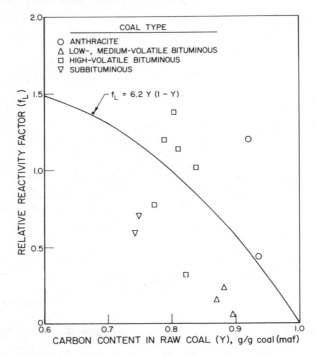

Figure 51 Comparison of relative reactivity factors from Tomita and colleaques with correlation proposed by Johnson. Data from Tomita, Mahajan, et al. [168] and Johnson [121].

a catalytic component in an oxidized state, gasification of the char in hydrogen could be accompanied by reduction of the oxide. If the reduced state of the catalyst is more effective than the the oxidized state, gasification rates will initially increase until reduction of the catalyst is complete. Conversely, if the oxidized state is more effective, then initial gasification rates will decrease.

Another explanation for decreasing gasification rates during the early reaction stages has been proposed by Blackwood [137, 167]. He suggests that when a coal char is prepared in an inert atmosphere, the char surface becomes saturated with $-CH_2$ complexes. Subsequently, when exposed to a hydrogen atmosphere, these complexes can react relatively rapidly with hydrogen, according to the reaction given by step 1 of the mechanism proposed by Blackwood and coworkers [137, 165-167] (Reaction 12), to produce a surge of methane, as the chemisorbed $-CH_2$ complexes approach the lower steady-state concentrations characteristic of reaction conditions. Blackwood suggests that the removal of excess $-CH_2$ complexes is rapid, occurring in less than a few minutes at practical char gasification temperatures above 700°C. Removal may even occur in seconds, since gas mixing in experimental reactors can mask the time effect of an impulse of methane. Blackwood suggests that this effect can also be seen in the data obtained by Moseley and Patterson [169] with bituminous coal char.

An additional explanation for the decreasing rates during early gasification stages is related to the rapid-rate methane reaction discussed previously. This can occur if the char-preparation conditions are such that complete deactivation of the intermediate responsible for rapid-rate reaction has not occurred. Conditions favoring this effect would be low temperatures and a short residence time during char formation.

Some of the transient effects that can result during coal-char gasificiation with hydrogen under unsteady-state conditions can be examined in more detail in terms of the mechanism described by steps 1 and 2 (Reactions 12 and 13). The steady-state gasification rate predicted by this mechanism is given by Equation 81, and the steady-state concentration of the chemisorbed $-CH_2$ complex $(n^*)_{ss}$ is given by

$$(n^*)_{ss} = \frac{(k_1 P_{H_2} + k'_2 P_{CH_4}) n_T}{k'_1 + (k_1 + k_2) P_{H_2} + k'_2 P_{CH_4}} \quad [97]$$

In the absence of methane, this simplifies to

$$(n^*)_{ss} = \frac{k_1 P_{H_2} n_T}{k'_1 + (k_1 + k_2) P_{H_2}} \quad [98]$$

and for the case where $k_2 \gg k_1$ and $k_2 P_{H_2} \gg k'_1$, then

$$(n^*)_{ss} = \frac{k_1}{k_2} n_T \quad [99]$$

Equation 98 leads to the steady-state rate expression given by Equation 78, and Equation 99 leads to the steady-state rate expression given by Equation 79.

When the $-CH_2$ surface complex concentration, n^*, is not at steady state under given conditions of temperature and hydrogen pressure, the following equation applies for gasification in hydrogen:

$$\frac{dn^*}{dt} = k_1 n_T P_{H_2} - [k'_1 + (k_1 + k_2) P_{H_2}] n^* \quad [100]$$

and when

$$k_2 \gg k_1 \text{ and } k_2 P_{H_2} \gg k'_1 \quad [101]$$

116 FUNDAMENTALS OF COAL GASIFICATION

$$\frac{dn^*}{dt} = k_1 n_T P_{H_2} - k_2 P_{H_2} n^* \qquad [102]$$

and

$$W = \frac{dX/dt}{1 - X} = k_2 P_{H_2} n^* \qquad [103]$$

Equations 102 and 103 bring up some interesting points. One involves the situation where a coal char is first exposed to a hydrogen atmosphere at a relatively low temperature such that steady-state gasification rates are negligible but the char surface can become saturated with respect to $-CH_2$ complex. The char is then directly exposed to a high-temperature hydrogen environment. Blackwood has experimentally shown that, with this type of treatment, exposure of the char to high temperatures results in an initial burst of methane formed at a rapid rate, followed by an approach to the steady-state rate characteristic of the temperature and pressure used [137]. This behavior can be explained in terms of Equations 102 and 103 as follows. Integrating Equation 102 at constant temperature and pressure yields

$$n^*(T, t) = \frac{k_1 n_T}{k_2} + \left\{ n_0^* - \left(\frac{k_1 n_T}{k_2} \right) \right\} e^{-k_2 P_{H_2} t} \qquad [104]$$

where n_0^* = concentration of $-CH_2$ complex at time $t = 0$. The steady-state concentration of $-CH_2$, $(n^*)_{ss}$, at temperature T is given by Equation 99. Thus

$$n^*(T, t) = (n^*)_{ss} + [n_0^* - n_{ss}^*] e^{-k_2 P_{H_2} t} \qquad [105]$$

Substituting Equation 105 into Equation 103 and integrating with respect to time yields

$$-\ln \frac{1 - X}{1 - X_0} = k_2 P_{H_2} \left[(n^*)_{ss} t + \frac{[n_0^* - n_{ss}^*]}{k_2 P_{H_2}} (1 - e)^{-k_2 P_{H_2} t} \right] \qquad [106]$$

or, at sufficiently long times, when $\exp(-k_2 P_{H_2} t) \cong 0$:

$$-\ln \frac{1 - X}{1 - X_0} = k_2 P_{H_2} [(n^*)_{ss} t] + [n_0^* - n_{ss}^*] \qquad [107]$$

Equation 107 shows an excess of conversion during time t, which is reflected by the term $(n_0^* - n_{ss}^*)$, relative to conversion that would be obtained under steady-state conditions, that is, $k_2 P_{H_2} n_{ss}^* t$, if $n_0^* > n_{ss}^*$. This condition of excess is satisfied by the initial treatment of the char at a lower temperature, since steady-state concentrations of the surface complex decrease with increasing temperature. Moseley and Patterson [169] have observed this "burst" effect, and Blackwood [137] has shown that the excess methane formed by the treatment described earlier in this paragraph can be continuously repeated.

If the char is cycled between two pressures at constant temperature, instead of being cycled between two temperatures at constant pressure, there are no transient effects in cases where Equation 99 can be applied, since n_{ss}^* depends only on temperature. Moseley and Patterson [169] observed this type of behavior when the hydrogen pressure is suddenly increased from 42 atm to 96 atm. Although the rate was higher at 96 atm, no transient effects occurred. For conditions under which Equation 98 is applicable, an increase in hydrogen pressure would result in an interval in which gasification rates increased to the steady-state value at the higher pressure, since $(n^*)_{ss}$ increases somewhat with increasing pressure. At pressures above around 10 atm, however, this effect would be small.

Dent [176] has shown that coal chars subjected to continuously increasing temperatures show a higher reactivity during hydrogen gasification than would be expected on the basis of rates obtained isothermally. This result can be predicted from Equations 102 and 103. If, for example, the concentration of the $-CH_2$ complex is at a steady-state value at some temperature T_0 and the temperature is then increased at a uniform rate, the value of n^* will always be greater than n_{ss}^* since n_{ss}^* decreases with increasing temperature. Thus the steady-state rate given by $W_{ss} = k_2 P_{H_2} n_{ss}^*$ will always be less than the actual rate, given by $W = k_2 P_{H_2} n^*$.

This example illustrates a possible danger in integrating rate expressions obtained under steady-state conditions to describe behavior under unsteady-state conditions. Conversely, with respect to the char–hydrogen reaction, it indicates that rate constants interpreted directly in terms of an expression such as Equation 79 from data obtained in experiments with constant heat-up rates could be too high. This would give an incorrect activation energy. This is one possible explanation for the relatively high rate constants reported by Jüntgen (see Figure 47), which were determined from experiments with constant heat-up rates. The high rate constants shown in Figure 47, deduced from Moseley's experiments under assumed steady-state conditions, may also be excessive because of the experimental procedure employed in cyclically heating and cooling the coal char in hydrogen.

Coal–Char–Carbon Dioxide–Carbon Monoxide System. The reaction of coal chars and other carbon solids with carbon dioxide:

$$CO_2 + C = 2CO \qquad [2]$$

has been studied for many years because of the interest in both gasification and combustion systems. Studies prior to the 1960s that have been reviewed by von Fredersdorff and Elliott [8] were largely conducted at atmospheric and subatmospheric pressures, using relatively purified carbon solids rather than actual coal chars. An excellent review of more recent studies has been prepared by Mentser and Ergun [177]. Although experimental work during the last 15 years has still focused primarily on the use of purified carbons at low pressure, increasing attention has been paid to gasification at elevated pressures.

Significant differences between the gasification of carbons with carbon dioxide and the gasification with hydrogen have been found. At atmospheric conditions, for example, gasification rates in pure carbon dioxide are generally greater than rates in pure hydrogen by a factor of 10^2 to 10^3. However, although gasification rates in hydrogen increase almost in direct proportion to pressure over a wide pressure range, rates in carbon dioxide have usually been found to increase with pressure only at relatively low pressure levels and then to plateau at pressures less than approximately 10 atm. Such an effect with carbon dioxide is reflected in the following kinetic correlation, which is frequently employed to describe the gasification kinetics of carbon in carbon dioxide and carbon monoxide mixtures, at conditions reasonably removed from equilibrium for this reaction:

$$W = \frac{dX/dt}{1-X} = \frac{k_c P_{CO_2}}{1 + K_{c,1} P_{CO} + K_{c,2} P_{CO_2}} \qquad [108]$$

where

$k_c, K_{c,1}, K_{c,2}$ = kinetic parameters that depend on temperature and nature of char

P_{CO}, P_{CO_2} = carbon monoxide and carbon dioxide partial pressures

Equation 108 shows that gasification rates decrease with increasing carbon monoxide partial pressure because of an inhibiting effect that is not directly related to thermodynamic reversibility. This contrasts to the behavior observed in carbon gasification with hydrogen, in which the effects of methane are usually reflected only in terms of thermodynamic reversibility and not as a poisoning agent. At a sufficiently high total pressure, Equation 108 simplifies to the following expression, which has been used in some studies to analyze experimental results:

$$W = \frac{k_c^*}{1 + K_c^* P_{CO}/P_{CO_2}} \qquad [109]$$

where

$$k_c^* = k_c/K_{c,2} \qquad [110]$$
$$K_c^* = K_{c,1}/K_{c,2} \qquad [111]$$

Although Equation 108 has been rationalized with a variety of mechanisms, the following reaction steps have been postulated most often in recent analyses [177]:

Step 5. $\qquad C_f + CO_2 \underset{k_5'}{\overset{k_5}{\rightleftarrows}} C(O) + CO \qquad [18]$

Step 6. $\qquad C(O) \underset{k_6'}{\overset{k_6}{\rightleftarrows}} CO + C_f \qquad [19]$

Step 5 (steps 1-4 were shown in Reactions 12-15) represents an oxygen-exchange reaction in which a carbon dioxide molecule dissociates at an active site, C_f, on the carbon surface, releasing a molecule of carbon monoxide and forming a solid carbon–oxygen complex, $C(O)$. This reaction, considered reversible, results only in the exchange of oxygen with the solid and does not result in the gasification of solid carbon. The actual carbon gasification occurs in step 6 (Reaction 19), with the dissociation of the carbon–oxygen surface complex from the bulk carbon matrix, leading to the formation of carbon monoxide and to the generation of a new active center on the solid carbon surface.

Assuming that the total number of active sites plus carbon–oxygen complexes is constant and assuming further that under steady-state conditions the net rate of formation of carbon–oxygen complex is zero, the following rate expression has been derived, based on steps 5 and 6 (Reactions 18 and 19):

$$W = \frac{k_5 n_t P_{CO_2} [1 - (k_5' k_6'/k_5 k_6)(P_{CO}^2/P_{CO_2})]}{1 + [(k_6'/k_6) + (k_5'/k_6)]P_{CO} + (k_3/k_6)P_{CO_2}} \qquad [112]$$

where n_t is the concentration of total free active sites plus oxygen–carbon complexes, in mol/mol solid carbon Equation 112 is analogous to Equation 108 with the identities

$$k_c = k_5 n_t \qquad [113]$$

$$K_{c,1} = \left(\frac{k_6'}{k_6} + \frac{k_5'}{k_6}\right) \qquad [114]$$

$$K_{c,2} = \frac{k_5}{k_6} \qquad [115]$$

120 FUNDAMENTALS OF COAL GASIFICATION

To satisfy thermodynamic equilibrium requirements, the term $(k'_5 k'_6 / k_5 k_6)$ in Equation 112 should be equal to the equilibrium constant, $K_{5,6}$, for the overall reaction given by Reaction 2 that is,

$$\frac{k_5 k_6}{k'_5 k'_6} = K_{5,6} \qquad [116]$$

This term is usually large enough, under the conditions employed in most experimental studies, that reversibility effects of carbon monoxide are swamped by the poisoning effect of carbon monoxide.

The parameters in the reduced form of Equation 109 are related to Equation 112 by the following expressions:

$$k_c^* = k_6 n_t \qquad [117]$$

$$K_c^* = \frac{k'_6}{k_5} + \frac{k'_5}{k_5} \qquad [118]$$

For the condition in which $k'_6 \ll k'_5$ (which often occurs), the term

$$K_c^* \cong \frac{k'_5}{k_5} \qquad [119]$$

represents the equilibrium constant for the oxygen-exchange reaction in step 5 (Reaction 18). The temperature coefficient for this term, when expressed by the Gibbs-Helmholtz equation, represents the heat of reaction for this step.

The rate expressions described by Equations 108 and 112 are based not only on results of studies that determined total gasification rate as a function of temperature, pressure, and gas composition, but also on studies using isotopic gases and/or solid carbons. The use of isotopes to study the mechanism of the carbon–carbon dioxide reaction has been a particularly useful tool, although not applicable to the study of the stoichiometrically simpler carbon–hydrogen system. The fact that similar evaluations of the oxygen-exchange equilibrium constant, $K_5 (= k_5/k'_5)$, have been made from indirect results obtained from kinetic measurements of the overall gasification reactions and by direct measurement using isotopic techniques is perhaps the strongest evidence for the validity of oxygen exchange as an essential reaction step.

Different investigators, however, have postulated other reaction steps in addition to or in place of steps 5 and 6 (Reactions 18 and 19). For example, Blackwood and Ingeme [178] have proposed a mechanism different from that expressed by steps 5 and 6 (Reactions 18 and 19) to account for results obtained on gasification of a coconut char in carbon dioxide–carbon monoxide mixtures, at pressures up to 40 atm. This study is particularly significant because it represents one of the few reported in the literature for carbon gasification with carbon dioxide at elevated pressures. Blackwood and Ingeme postulated the following sequence:

Step 5. (forward) $\qquad C_f + CO_2 \xrightarrow{k_5} CO + C(O)$ [18]

Step 6. (forward) $\qquad C(O) \xrightarrow{k_6} CO + C_f$ [19]

Step 7. $\qquad CO + C_f \underset{k_7'}{\overset{k_7}{\rightleftarrows}} C(CO)$ [20]

Step 8. (forward) $\qquad CO_2 + C(CO) \xrightarrow{k_8} 2CO + C(O)$ [21]

Step 9. (forward) $\qquad CO + C(CO) \xrightarrow{k_9} CO_2 + 2C_f$ [22]

Although over the years a number of other investigations have also postulated mechanisms for carbon gasification by carbon dioxide that include reaction steps involving chemisorbed carbon monoxide, such studies represent a minority view. As discussed recently by Mentser and Ergun [177], there are many more studies that have provided various types of evidence to support the thesis that carbon monoxide retards reaction, not through chemisorption, but by the reverse oxygen-exchange reaction [step 5 (Reaction 18), reverse), which decreases the number of surface oxygen complexes that can react by step 6 (Reaction 19, forward) to gasify carbon.

Perhaps the most significant part of the mechanism proposed by Blackwood and Ingeme is in the rate equation derived from it which describes results obtained at elevated pressures. Based on a number of simplifying approximations, the following equation was proposed:

$$W = \frac{(k_5 P_{CO_2} + (k_5 k_8/k_7') P_{CO_2}^2) n_t}{1 + (k_7/k_7') P_{CO} + (k_5/k_6) P_{CO_2}} \qquad [120]$$

With the exception of the term $P_{CO_2}^2$ in the numerator of Equation 120, the form of expression is the same as that of Equation 108, which is frequently used to describe results of atmospheric-pressure studies. As shown by the numerical evaluation of Blackwood and Ingeme's results obtained on gasification of coconut charcoal at 790–870°C and a total pressure up to 40 atm (Table 17), the term $(k_5 k_8/k_7') P_{CO_2}^2 n_t$ can be neglected at carbon dioxide partial pressures less than 1 atm. At sufficiently high pressures, however, Equation 120 predicts that,

TABLE 17 NUMERICAL VALUES FOR CONSTANTS OF THE RATE EQUATION[a]

Temperature (°C)	$k_5 n_t$ (min^{-1} atm^{-1} × 10^4)	$\left(\dfrac{k_5 k_8}{k_7'}\right) n_t$ (min^{-1} atm^{-2} × 10^4)	$\dfrac{k_7}{k_7'}$ (atm^{-1})	$\dfrac{k_5}{k_6}$ atm^{-1}
790	0.62	0.048	13.7	0.23
830	2.76	0.24	17.1	0.17
870	7.80	0.84	12.7	0.31

[a] Data from Blackwood, Ingeme, et al. [178].

at constant gas composition, gasification rates will become proportional to total pressure, that is,

$$W \cong \frac{n_t(k_5 k_8/k_7')P_{CO_2}^2}{(k_7/k_7')P_{CO} + (k_5/k_6)P_{CO_2}} \quad [121]$$

rather than becoming independent of total pressure, as predicted by Equation 108. Unfortunately, little evidence is available in the literature to support the large effects of pressure predicted by Blackwood and Ingeme's correlation, particularly at elevated pressures.

In an unpublished study conducted at the IGT, results obtained for the gasification of bituminous coal char in carbon monoxide and carbon dioxide mixtures, at 2-35 atm and 850-1000°C, were correlated with an expression of the form given by Equation 112. That equation predicts that gasification rates will eventually become independent of pressure, under conditions reasonably removed from equilibrium. The IGT study made the following evaluations, based on gasification of an hvAb coal char prepared by devolatilizing the raw coal is nitrogen at the same temperature used for subsequent gasification:

$$k_c = k_5 n_t = 3.56 \times 10^3 \exp\left(\frac{-28,430}{RT}\right) \quad [122]$$

$$K_{c,1} = \frac{k_6'}{k_6} + \frac{k_5'}{k_6} = 0.15 \exp\left(\frac{6,400}{RT}\right) \quad [123]$$

$$K_{c,2} = \frac{k_5}{k_6} = 1.04 \times 10^{-7} \exp\left(\frac{36,500}{RT}\right) \quad [124]$$

where k_c is in min^{-1} atm^{-1}, $K_{c,1}$ is in atm^{-1}, and $K_{c,2}$ is in atm^{-1}.

The effects of total pressure on gasification rates as predicted by Blackwood and Ingeme's correlation [178] are contrasted in Table 18 to those determined

TABLE 18 COMPARISON OF PRESSURE EFFECTS ON GASIFICATION RATES[a]

Pressure (P) (atm)	CO_2 Only		10% CO, 90% CO_2	
	IGT	Ref. 178	IGT	Ref. 178
1	1.00	1.00	1.00	1.00
10	1.80	5.98	1.71	2.78
30	1.91	14.56	1.80	5.75
∞	1.98	∞	1.90	∞

[a] Specific rate ratio, $W(P)/W(1\ \text{atm})$.

in the IGT study, which involved gasification in carbon dioxide and in a carbon dioxide and a carbon monoxide mixture at 870°C. Although Blackwood and Ingeme's correlation predicts continually increasing rates with increasing pressure, the IGT correlation indicates that total pressure has relatively little effect at pressures greater than 10 atm. A trend similar to that suggested by the IGT correlation was also reported by Fuchs and Yavorsky [179], who studied the kinetics of the gasification of bituminous coal chars in carbon dioxide under conditions of differential gas conversion. These investigators found that, at 750°C and 800°C, total pressure had no effect on gasification rates at pressures of 18-35 atm. Since it seems unreasonable to attribute the significant differences in the effects of total pressure solely to the different carbon solids used in the few studies conducted at elevated pressure, it appears that additional research would be useful in elucidating the mechanism of the carbon (coal char)-carbon dioxide reaction at high pressures.

Quite recently, Yergey and Lampe [180] also proposed a reaction mechanism different from that expressed by steps 5 and 6 (Reactions 18 and 19), based on experimental results observed for the gasification of an isotopic mixture of purified carbons (^{12}C and ^{13}C) with carbon dioxide ($^{12}CO_2$):

Step 10. $$C_f + CO_2 \underset{k'_{10}}{\overset{k_{10}}{\rightleftarrows}} C(CO_2) \qquad [23]$$

Step 11. $$C(CO_2) \underset{k'_{11}}{\overset{k_{11}}{\rightleftarrows}} 2CO + C_f \qquad [24]$$

Marsh and Taylor [181], however, subsequently argued that the conclusions made by Yergey and Lampe were invalid because they misinterpreted their experimental results, primarily because they used a solid carbon that was not degassed prior to reaction with carbon dioxide. It should also be noted that reaction steps 10 and 11 (Reactions 23 and 24) do not lead to a kinetic expres-

124 FUNDAMENTALS OF COAL GASIFICATION

sion of the form given by Equation 115 or 119, but instead lead to an expression with a denominator containing a P_{CO}^2 term in place of a P_{CO} term, contrary to the vast majority of kinetic studies made.

Since the existing evidence favors the mechanism given by steps 5 and 6 (Reactions 18 and 19), it is of interest to compare the numerical evaluations of kinetic parameters determined from data obtained by different investigators (See Table 19). Such a comparison, in terms of the notation given in Equation 112, is shown in Table 20. Results in Tables 19 and 20 were prepared with the common assumption that under the reaction conditions experimentally employed, the

TABLE 19 COMPARISON OF KINETIC PARAMETERS FROM DIFFERENT INVESTIGATORS

Reference	Curve (Figures 52-54)	Carbon Type	Temperature (°C)	Pressure (atm)
182	A	Subbituminous coal char	600-790	1
179	B	High-volatility bituminous coal char (Illinois seam)	750-900	18-35
183	C_1	High-volatility bituminous coal char (Pittsburgh seam)	880-1080	1
183	C_2	High-volatility bituminous coal char (Illinois seam)	850-1080	1
(IGT)	D	High-volatility bituminous coal char (Pittsburgh seam)	870-980	1-35
177	E	Spheron No. 6 carbon	750-850	1
184	F	Graphite	–	–
185	G	Coke	–	–
186	H	SP-1 Graphite	1050-1200	1
187	I	SP-1 Graphite	900-1000	1
188	J	Coke	900-1000 1000-1200	1
178	K	Coconut charcoal	790-870	2-40
189	L_1	Activated carbon	700-1000	1
189	L_2	Activated graphite	900-1150	1
189	L_3	Ceylon graphite	1000-1400	1
190	M_1	Coke	800-1100	1
190	M_2	Anthracite	800-1100	1
191	N	Coconut charcoal	730-830	1

TABLE 20 COMPARISON OF KINETIC PARAMETERS USING DIFFERENT EQUATION NOTATIONS

Curve[a]	$k_5 n_t = A_5 \exp(-E_5/RT)$		$k_5' n_t = A_5' \exp(-E_5'/RT)$		$k_6 n_t = A_6 \exp(-E_6/RT)$		$K_5 = \dfrac{k_5}{k_5'} = \dfrac{A_5}{A_5'} \exp[-(E_5 - E_5')/RT]$	
	A_5 (min^{-1} atm^{-1})	E_5 (kcal/g-mol)	A_5' (min^{-1} atm^{-1})	E_5' (kcal/g-mol)	A_6 (min^{-1})	E_6 (kcal/g-mol)	A_5/A_5'	$E_5 - E_5'$ (kcal/g-mol)
A	—	—	—	—	—	—	—	—
B	—	—	—	—	6.8×10^4	54.5	—	—
C_1	—	—	—	—	—	—	—	—
C_2	—	—	—	—	—	—	—	—
D	3.56×10^3	28.4	5.02×10^9	58.6	3.42×10^{10}	65.0	7.09×10^{-7}	−30.2
E	1.54×10^6	53.0	4.48×10^3	36.0	5.15×10^7	58.0	3.44×10^2	17.0
F	—	41	—	—	—	124	—	—
G	—	61.7	—	—	—	67.8	—	34.2
H	—	103.5	—	27.5	—	72.5	—	24.0
I	—	99.0	—	79.5	—	87.0	—	25.0
J	—	—	—	74.0	—	65.3	1.72×10^4	27.0
K	2.7×10^{11}	76.0	1.63×10^{13}	76.0	8.82×10^8	31.5	—	—
L_1	—	—	—	—	1.43×10^3	76.0	1.66×10^{-2}	0.0
L_2	—	—	—	—	1.13×10^{12}	59.0	4.15×10^3	23.0
L_3	—	—	—	—	8.58×10^8	59.0	4.15×10^3	23.0
M_1	8.3×10^6	47.6	5.52×10^5	38.9	8.28×10^7	59.0	4.15×10^3	23.0
M_2	2.6×10^4	32.5	6.76×10^4	32.2	1.55×10^7	53.9	15.0	8.7
					3.95×10^7	49.1	0.39	0.3
N	7.6×10^9	58.8	3.01×10^{-5}	−16.8	1.47×10^7	28.7	2.62×10^{14}	75.6
					2.39×10^3			
Median Value	—	53	—	39–59	—	59	—	23

[a] Curves correspond to same literature sources as those in Table 19.

rate of step 6 (Reaction 19, reverse) is slow compared to step 5 (Reaction 18, reverse). Thus

$$\frac{k'_5}{k_6} \ll \frac{k'_6}{k_6} \qquad [125]$$

and, from Equation 115,

$$K_{c,1} = \frac{k'_6}{k_6} \qquad [126]$$

The results in Tables 19 and 20 include the evaluation of earlier work that was summarized by von Fredersdorff and Elliott [8] and some of the more recent results summarized by Mentser and Ergun [177]. The values of kinetic parameters shown were either reported directly by individual investigators or were estimated from reported, graphic, or tabular information reduced to a common basis. As pointed out by Mentser and Ergun, individual rate constants cannot be determined directly from reported data without knowing the concentration of active sites, n_t. Thus the apparent activation energies of the individual terms may be affected by changes in the concentration of active sites with changing temperature. As can be seen, these studies show a relatively wide range of values of parameters, as shown in Equation 112. This is partly due, of course, to the different carbonaceous solids used and to the different experimental procedures and methods of data analyses employed.

Graphic representation of the terms K_5 and $k_6 n_t$ are shown in Figures 52 and 53, respectively. In spite of the range of values for differences in activation energy $(E_5 - E'_5)$ for the parameter K_5, the values of K_5 predicted from the correlation obtained using different carbons show a surprisingly similar overall trend (Figure 52), reasonably well supported by evaluations made from Ergun's data [189], that is, $K_5 = 4.15 \times 10^{13} \exp(-23{,}000/RT)$.

This similarity gives some credence to the supposition that K_5 is in fact an equilibrium constant representative of the oxygen-exchange reaction in step 5 (Reaction 18), and is relatively independent of carbon type.

If this is true, the values of $E_5 - E'_5 \cong 23$ kcal/g-mol can be considered as the heat of reaction, ΔH, for the oxygen-exchange reaction in step 5 (Reaction 18). This magnitude of ΔH has been interpreted by some investigators to imply certain characteristics of the nature of carbon–oxygen bonding in the solid complex, -C(O), formed in the oxygen-exchange reaction. According to Mentser and Ergun [177], for example, the probable values of ΔH obtained in recent investigations suggest that the bonding between oxygen and carbon in the surface complex is, on the average, intermediate between that of true single bonds and double bonds, although somewhat closer to double bonds. Other investigators have reached similar conclusions.

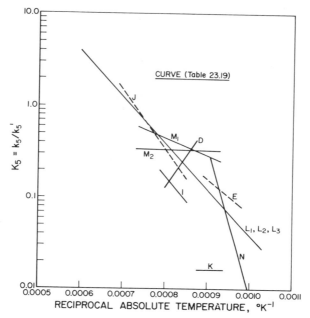

Figure 52 Oxygen-exchange equilibrium constant as determined by various investigators.

Values of $k_6 n_t$, as shown in Figure 53, do exhibit significant differences for the various carbon solids. Although the overall activation energies for different materials are comparable, clustered about a median value of 59 kcal/g-mol, the preexponential factors vary by a factor of about 100-200. In general, the variations in the term $k_6 n_t$ are due to possible variations in both k_6 and n_t. If, however, it is assumed that k_6 is a fundamental rate constant, similar for different carbon solids, major differences in the magnitude of values of $k_6 n_t$ for different materials can be attributed to differences in values of the concentration of active sites, n_t. The value of this term would be expected to decrease generally with increasing solid preparation or reaction temperature for any material, as suggested previously in the discussion of the carbon–hydrogen reaction and also as indicated by Duval's theory of the neutralization of active sites in carbon solids [192, 193]. For coal chars, n_t would also be expected to decrease with increasing coal rank. These general trends are reasonably substantiated by the approximate order of the values of $k_6 n_t$ exhibited in Figure 53 for curves with similar slopes, that is,

$$D > M_2, L_1 > K, M_1 > L_2, J, E > L_3 \qquad [127]$$

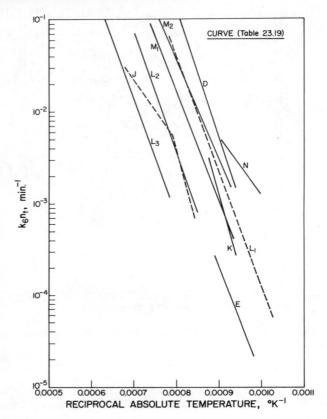

Figure 53 Composite rate constant compiled from various investigations.

The sequence for the carbon solids referred to by each curve is: high-volatility bituminous coal char (Pittsburgh seam) > anthracite, activated carbon > coconut charcoal, coke (M_1) > activated graphite, coke (J), spheron 6 carbon black > Ceylon graphite.

Another way to consider the relative reactivities of different carbon solids in the carbon–carbon dioxide reaction is to compare gasification rates under a standard condition or range of conditions. This type of procedure in part avoids some of the problems associated with the accurate evaluation of individual kinetic parameters and reflects overall reaction rate measurements to a greater extent. Values of the specific reaction rate, W, are given in Figure 54 as a function of the temperature at 1 atm, in pure carbon dioxide, based on the use of kinetic parameters given in Tables 19 and 20 for the different carbon solids. Figure 54 also includes rate values computed from two investigations with coal chars, in which results were correlated with empirical rate expressions that dif-

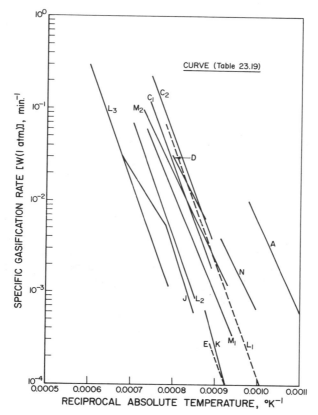

Figure 54 Gasification rates in pure CO_2 at 1 atm predicted from correlations of various investigations.

fered from the form given by Equation 119. Each of these studies was conducted at atmospheric pressure with pure CO_2, under conditions in which carbon monoxide concentrations were very small. In one study conducted with subbituminous coal char, reported by Taylor and Bowen (curve A) [182], specific gasification rates were correlated with the expression

$$W = 4.2 \times 10^7 \exp\left(\frac{-45{,}000}{RT}\right) \qquad [128]$$

(W in min^{-1}) for temperatures of 600–790°C, under conditions in which the chars were produced at reaction temperature. Taylor and Bowen report that specific gasification rates for chars prepared from 6-mm-diameter subbituminous coal are relatively constant, at least for char conversions up to 90%.

Dutta, Wen, et al. [183] expressed gasification rates of chars derived from high-volatility bituminous coals in carbon dioxide, at 1 atm and 850-1080°C, with a correlation of the form

$$W = ak_v C \qquad [129]$$

where

k_v = rate constant

C = carbon dioxide concentration

a = parameter that reflects the relative available internal char surface area

This correlation was used to describe rates obtained with 35 + 60-mesh chars obtained from Illinois and Pittsburgh seam coals. Parameter a decreased with carbon conversion for some of the chars tested (particularly those from the Pittsburgh seam coal) and can be considered to be proportional to the total concentration of active sites, n, defined previously. Thus variations in a can be due both to a variation in the available char surface area and a variation in the active sites per unit of available surface area. Variations in available surface area would be more likely to occur during gasification under isothermal conditions. For gasification at 1 atm of total pressure, initial gasification rates, where $a = 1$ by definition, can be expressed by

$$W = \frac{k_v}{RT} \qquad [130]$$

based on Equation 129. Average values of kinetic parameters for the term k_v/RT, for the two types of char used by Dutta, can be deduced from the following reported information:

Pittsburgh seam:

$$W = \frac{k_v}{RT} = (0.61 \pm 0.13) \times 10^9 \exp\left(\frac{-59,260}{RT}\right) \qquad [131]$$

Illinois seam:

$$W = \frac{k_v}{RT} = (1.21 \pm 0.25) \times 10^9 \exp\left(\frac{-59,260}{RT}\right) \qquad [132]$$

where W is in min^{-1}. It should be noted that the deviations indicated previously are not based on reported evaluations, but were derived by averaging reported parameters for different samples and using the single reported activation energy, based on concentration units, to express the deviations in units of pressure, over the range of temperatures employed.

Evaluations of specific gasification rates at 1 atm are shown in Figure 54 as a function of temperature. Similar to the trend exhibited by values of $k_6 n_t$, values of W generally show a wide variation, but with less of a specific trend for medium-reactivity carbons. It is convenient to evaluate the order of carbon reactivities by comparing their interpolated or extrapolated rates at 900°C (Table 21), although too great an extrapolation beyond reported temperature ranges may be meaningless.

The values given in Table 21 again show the expected trends, with some exceptions. The gasification rate for subbituminous char (A) appears to be unusually high, indicating the possibility of catalysis or initial-stage reaction effects, since actual rate measurements were made at relatively low temperatures. Also, the rates for the two coconut charcoals (N and K) are different for reasons that are not easily apparent. The higher value (N, 0.012 min^{-1}) is comparable to values obtained with bituminous coals, whereas the low value (K, 0.002 min^{-1}) is comparable to values obtained with high-temperature coke. This low value may be particularly suspect, in view of the fact that Blackwood and Ingeme's correlation [178], used to evaluate the low rate, predicts gasification rates at higher pressures that are comparable to those of bituminous coals. This is illustrated in Figure 55, which also includes rates reported by Fuchs and Yavorsky [148] for the gasification of high-volatility bituminous coal char in pure carbon dioxide at 18-35 atm and 750-900°C.

TABLE 21 COMPARISON OF SPECIFIC GASIFICATION RATES FROM FIGURE 54 (900°C)

Curve[a]	Carbon Type	Rate (W) (min^{-1})
A	Subbituminous coal char	0.23
C_2	High-volatility bituminous coal char (Illinois)	0.012
N	Coconut char	0.012
D	High-volatility bituminous coal char (Pittsburgh)	0.011
L_1	Activated carbon	0.009
M_2	Anthracite	0.007
C_1	High-volatility bituminous coal char (Pittsburgh)	0.006
M_1	Coke	0.003
K	Coconut char	0.002
L_2	Activated graphite	0.0009
E	Carbon black	0.0009
J	Coke	0.0006
L_3	Ceylon graphite	0.0002

[a]Curves correspond to same literature sources as those in Table 19.

132 FUNDAMENTALS OF COAL GASIFICATION

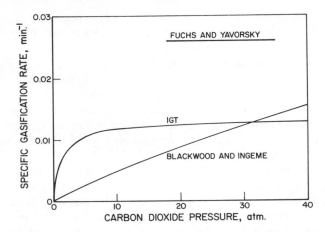

Figure 55 Effect of pressure on gasification in pure CO_2 at $870°C$. Data from Fuchs and Yavorsky [179] and Blackwood and Ingeme [178].

Specific gasification rates in carbon dioxide at 1 atm and 900°C have also been determined in studies conducted at Pennsylvania State University [99, 194, 195] for chars derived from coals ranging widely in rank from anthracite to lignite. The selection of these values in Table 22 shows that gasification rates can be reasonably correlated with rank, which is not inconsistent with the results given in Table 21.

Figure 56 plots the rates given in Table 22 as a function of carbon content in the raw coals. Results for lignites are not included because of the strong possibility of significant, inherent catalytic effects with these materials. The curve drawn in Figure 56 represents the relative-reactivity-factor correlation developed by Johnson [122] to describe reactivity variations in the gasification of coal chars with hydrogen (Equation 74). In spite of the scatter of the data apparent in Figure 56, the results do suggest some relationship between active sites, which is important in the carbon–hydrogen reaction and the carbon dioxide–carbon reaction. This relationship is also indicated from a study reported by Blackwood [196] and by Blackwood and Ingeme [178], who examined the effect of char-preparation temperature on wood-char gasification rates obtained in carbon dioxide at 1.2 atm and 650°C. Results from this study, showing the effects of char-preparation temperatures of 750–1150°C on gasification rates, are illustrated in Figure 57.

The curve in Figure 57, which is reproduced from Figure 49, describes the relative effects of pretreatment temperatures on char–hydrogen gasification rates, based on the analysis of Blackwood, Cullis, et al. [166] and Johnson's corrections [7] and interpreted in terms of the effect of increasing tempera-

TABLE 22 SPECIFIC GASIFICATION RATES OF COAL CHARS IN CARBON DIOXIDE[a]

Coal Type	Carbon Content in Raw Coal (maf) (%)	Specific Gasification Rate (min^{-1})
Lignite (N. Dak.)	63.3	0.060–0.093
Lignite (Mont.)	70.7	0.058–0.105
Lignite (N. Dak.)	71.2	0.045–0.195
Lignite (Tex.)	74.3	0.020–0.032
Sbb A (Wyo.)	74.3	0.022–0.028
Sbb C (Wyo.)	74.8	0.055–0.077
Hv B (Ill.)	77.3	0.003–0.025
Hv C (Ill.)	78.8	0.010–0.011
Hv B (Ill.)	80.1	0.006–0.022
Hv C (Ind.)	81.3	0.010
Hv A (W. Va.)	82.3	0.0013–0.0032
Hv A (Ky.)	83.8	0.0033–0.0072
Lv (Pa.)	88.2	0.0012–0.0037
Anthracite (Pa.)	91.9	0.0022–0.0046

[a] At 1 atm and 900°C (40 × 100 mesh chars prepared in nitrogen at 1000°C).

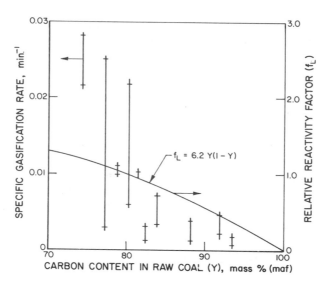

Figure 56 Effect of coal rank on gasification rate in CO_2 at 1 atm and 900°C. Data from Spackman, Davis, et al. [194] and Hippo and Walker [195].

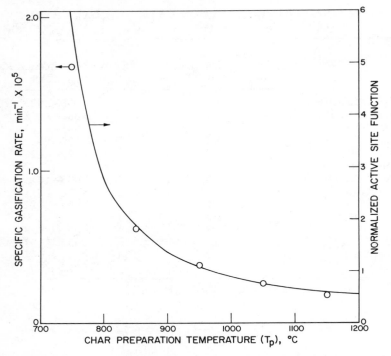

Figure 57 Effect of preparation temperature on gasification rates of wood chars in CO_2 at 1.2 atm and $650°C$. Data from Blackwood and Igeme [178].

ture, on decreasing active site concentration in the char–hydrogen reaction. The close correspondence of this curve to the data, however, strongly corroborates the evidence discussed previously that there is a direct relationship between the active sites involved in the carbon–hydrogen reaction and in the carbon–carbon dioxide reaction, although it does not prove that the sites participating in these two reactions are identical.

As indicated by Equation 112, an increase in the carbon monoxide partial pressure decreases the gasification rate. The relative decrease in gasification rates with increasing carbon monoxide mole fraction, z, at constant temperature and at total pressure, P, can be expressed by the following relationship:

$$\frac{W(z)}{W(z=0)} = \frac{1}{1 + (k_5'/k_6)zP/1 + (k_5/k_6)(1-z)P} \quad [133]$$

Quantitative examples of the relative effects of the carbon monoxide mole fraction on gasification rates at 1 atm pressure and 800–900°C are shown in Figures

58 and 59. These examples were calculated from Equation 139, using selected data given in Tables 19 and 20. Although the use of different kinetic parameters results in differences in the magnitude of the relative decrease in gasification rates with increasing temperature, the results of most studies predict that the effect of carbon monoxide will diminish with increasing gasification temperature. The general requirements for this trend to occur can be deduced by examining Equation 11. Under conditions in which $E_6 > E_5 > E'_5$, the denominator term in Equation 11 will always increase with increasing temperature, which results in decreasing relative gasification rates. This will also occur at pressures other than 1 atm. The above inequality is satisfied for results given in Tables 19 and 20 for cases E, G, M_1, and M_2. For most other cases in Tables 19 and 20, the situation is more complex since relative gasification rates theoretically can increase or decrease with increasing temperature, depending on pressure, gas composition, and temperature level. The one exception to this is case K, for which relative gasification rates do not change with increasing temperature. For conditions under which $k_5(1-z)P \gg k_6$, relative gasification rates at a specific carbon monoxide concentration increase with increasing temperature when

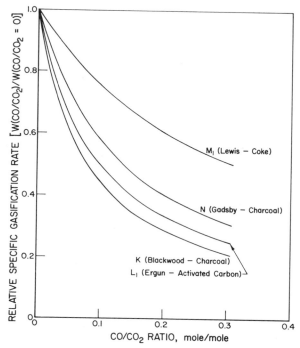

Figure 58 Effect of carbon monoxide concentration on gasification rates at 1 atm and 300°C.

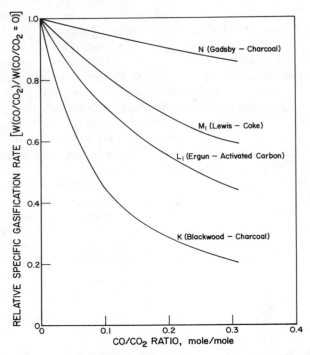

Figure 59 Effect of carbon monoxide concentration on gasification rates at 1 atm and 900°C.

$E_5 > E'_5$. This condition exists for almost all of the cases given in Tables 19 and 20.

Many carbonaceous materials, particularly coal chars, have some hydrogen in the solids that can gasify to form molecular hydrogen in the gas phase. The effect of hydrogen on the carbon–carbon dioxide reaction is thus of interest in practical gasification systems, even where carbon dioxide alone is employed as the feed gas. Unfortunately, relatively little work has been done to quantitatively define the effects of hydrogen concentration on this reaction, and the qualitative conclusions from available work are not consistent. Whereas investigations by Long and Sykes [197], Gadsby, Long, et al. [191], and Blakely and Overholser [198] have indicated inhibition due to hydrogen at atmospheric pressure, Blackwood and Ingeme [178] found that 10% hydrogen at 10 atm and 20 atm had no effect. Although von Fredersdorff and Elliott [8] have suggested that the different results may be due to differences in the pressure levels, it is apparent that this area needs further investigation.

Equation 112 was derived using reaction steps 5 and 6 (Reactions 18 and 19) for steady-state conditions. Certain types of behavior under nonsteady-state

KINETICS OF GASIFICATION REACTIONS   137

conditions can be predicted in a manner similar to that discussed previously with respect to the carbon–hydrogen reaction. One example of a nonsteady-state effect is that in which a carbon sample undergoing steady-state gasification in a carbon dioxide–carbon monoxide mixture is instantaneously exposed to a higher or lower temperature. For this case, an equation similar to Equation 107 can be derived to describe the relationship between carbon conversion and the amount of time after steady state is achieved at the new temperature:

$$-\ln \frac{1-X}{1-X_0} = k_6 n_{ss}^{**} t + \frac{k_6(n_0^{**} - n_{ss}^{**})}{k_5 f_{CO_2} + k_5' P_{CO} + k_6} \quad [134]$$

where n_{ss}^{**} is the steady-state concentration of the oxygen complex at the new temperature

$$\frac{n_t k_5 P_{CO_2}}{k_6 + k_5 P_{CO_2} + k_5' P_{CO}} = \frac{n_T (k_5/k_6) P_{CO_2}}{1 + (k_5/k_6) P_{CO_2} + (k_5'/k_6) P_{CO}} \quad [135]$$

and n_0^{**} is the steady-state concentration of the oxygen complex at the initial temperature.

It is apparent from Equation 134 that at the new temperature there will be either an excess or deficiency in carbon conversion relative to that predicted by applying the steady-state rate equation, depending on the sign of the term $(n_0^{**} - n_{ss}^{**})$. For conditions under which the steady-state concentration of the oxygen complex increases with increasing temperature, a temperature increase results in a deficiency in conversion, and a temperature decrease results in an excess of carbon conversion. This is the opposite of what was predicted for the carbon–hydrogen reaction. If the steady-state concentration of the oxygen complex decreases with increasing temperature, a behavior similar to that predicted for the carbon–hydrogen reaction will occur.

Coal Char–Synthesis Gas System. There are many complexities involved in the determination of reaction kinetics for the gasification of carbonaceous solids in systems containing the components carbon, hydrogen, and oxygen. This is mainly because the reaction stoichiometry in such systems is usually a function of reaction conditions rather than being approximately fixed, as is the case in the relatively simpler carbon–oxygen and carbon–hydrogen systems. Under some limiting conditions, reaction stoichiometry in carbon–hydrogen–oxygen systems is simplified because of the nature of the relative kinetics of the gasification reaction that occurs. At lower pressures, for example, methane formation is inhibited, and gasification occurs by some combination of Reactions 2, 3, and 10. In addition, at lower pressures—where steam and hydrogen alone

138 FUNDAMENTALS OF COAL GASIFICATION

predominate in the gas—the only reaction that occurs is that given by Reaction 10, and reaction stoichiometry is fixed.

At elevated pressures, however, methane becomes an increasingly important gasification product through overall reactions of the type in Reaction 1 and

$$H_2 + H_2O + 2C \rightleftharpoons CO + CH_4 \qquad [25]$$

Under such conditions, in which both methane and carbon oxides are the primary products of gasification, there are greater variations in the models or mechanisms proposed by different investigators (as well as the kinetic evaluations pertinent to these models) compared to evaluations made for the simpler gasification systems. In this sense, the few kinetic correlations developed to describe carbon gasification kinetics at elevated pressures with synthesis gases, or even with steam and hydrogen alone, are largely empirical in nature, with relatively little theoretical basis from a fundamental, mechanistic point of view. This does not, of course, preclude the practical application of these correlations to gasifier design, but does caution their use for reaction conditions not verified experimentally.

A wide variety of rate expressions has been used to correlate the gasification rates of carbon to form carbon oxides in steam–hydrogen mixtures, under various reaction conditions. Some of these are as follows:

$$W = k_{OH} P_{H_2O} \qquad [136]$$

$$W = \frac{k_{OH} P_{H_2O}}{1 + K_{OH,1} P_{H_2} + K_{OH,2} P_{H_2O}} \qquad [137]$$

$$W = k_{OH} \left(\frac{P_{H_2O}}{P_{H_2}}\right)^{1/2} \qquad [138]$$

$$W = \frac{k_{OH} P_{H_2O}^2}{1 + K_{OH,1} P_{H_2} + K_{OH,2} P_{H_2O}} \qquad [139]$$

$$W = \frac{k_{OH} P_{H_2O}}{(1 + K_{OH,1} P_{H_2} + K_{OH,2} P_{H_2O})^2} \qquad [140]$$

$$W = \frac{k_{OH} P_{H_2O} + k'_{OH} P_{H_2} P_{H_2O} + k''_{OH} P_{H_2O}^2}{1 + K_{OH,1} P_{H_2} + K_{OH,2} P_{H_2O}} \qquad [141]$$

where

$k_{OH}, k'_{OH}, k''_{OH}, K_{OH,1}, K_{OH,2}$ = kinetic parameters

P_{H_2}, P_{H_2O} = hydrogen and steam partial pressures

KINETICS OF GASIFICATION REACTIONS   139

With the exception of Equation 136, all the preceding rate expressions predict that hydrogen will inhibit gasification rates. Although some investigators have proposed Equation 136 on the basis of the limited available kinetic evidence, this equation can be applied only over a very narrow range of temperature, pressure, and gas composition.

Equation 137 has been used by several investigators to describe the kinetics of gasification at atmospheric and subatmospheric pressures over a wide range of temperatures. The studies using this expression for the gasification of cokes, graphites, and charcoals were made primarily in earlier investigations and have been reviewed by von Fredersdorff and Elliott [8]. They have shown, however, that kinetic parameters determined in different studies can show a wide variation. For the studies considered, the activation energy for the k_{OH} term was found to vary from 13 to 62 kcal/g-mol; for $K_{OH,1}$, from 0 to -75 kcal/g-mol; and for $K_{OH,2}$, from 20 to -83 kcal/g-mol. Despite these variations, which result from the strong covariance existing in quantitative evaluations of individual kinetic parameters, actual rate determinations show more consistent trends. This is illustrated in Figure 60 [199-201], which shows the rates of gasification in pure steam predicted by several correlations, at a pressure of 1 atm. Included in Figure 60 are more recent results obtained for the gasification of coal chars, as well as results of earlier studies with less reactive carbons.

One of the mechanisms proposed to describe the kinetics of the steam-carbon reaction is based on adsorption of steam and hydrogen on active sites [199, 202]. This mechanism assumes the following sequence:

Step 13. $\quad H_2O + C_f \underset{k'_{13}}{\overset{k_{13}}{\rightleftarrows}} C(H_2O)$ [26]

Step 14. $\quad C(H_2O) \underset{k'_{14}}{\overset{k_{14}}{\rightleftarrows}} CO + H_2 + C_f$ [27]

Step 1. $\quad H_2 + C_f \underset{k'_1}{\overset{k_1}{\rightleftarrows}} C(H_2)$ [12]

In the presence of hydrogen and steam alone, this mechanism leads to a rate expression of the form given in Equation 137, where

$$k_{OH} = \frac{k_{13}k_{14}n_t}{k'_{13} + k_{14}}$$ [142]

$$K_{OH,1} = \frac{k_1}{k'_1}$$ [143]

$$K_{OH,2} = \frac{k_{13}}{k'_{13} = k_{14}}$$ [144]

140 FUNDAMENTALS OF COAL GASIFICATION

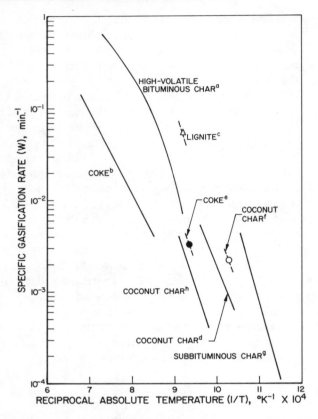

Figure 60 Comparison of carbon gasification rates in pure steam an 1 atm. Data from (a) Johnson [7], (b) Ergun [188], (c) Curran, Fink, et al. [201], (d) Long and Sykes [197], (e, f) Gadsby, Hinshelwood, et al. [199], and (g) Taylor and Bowen [182].

Long and Sykes [197] have proposed another mechanism that they interpret to lead to Equation 137. They suggest that steam adsorption is accompanied by dissociation into hydroxyl radicals and hydrogen atoms, followed by a surface reaction involving hydrogen exchange. The following sequence was proposed to replace steps 13 and 14 (Reactions 26 and 27):

Step 15. $\qquad H_2O + C_f \text{ (dual site)} \underset{k'_{15}}{\overset{k_{15}}{\rightleftarrows}} C(H)(OH)$ [28]

Step 16. $\qquad C(H)(OH) \underset{k'_{16}}{\overset{k_{16}}{\rightleftarrows}} C(H_2)(O)$ [29]

Step 6. $\qquad C(O) \underset{k_6'}{\overset{k_6}{\rightleftarrows}} CO + C_f \qquad$ [19]

This sequence, together with step 1 (Reaction 12), can lead to Equation 137 if it is assumed that step 16 (Reaction 29) occurs only in the forward direction and that the concentration of dual sites interacting in step 15 (Reaction 28) is equal to or directly proportional to the total concentration of free active sites. Although the fundamental significance of these assumptions is not readily apparent, the fact that individual rate steps depend only on first-order surface-species concentrations makes it much easier to derive the overall gasification rate expression. Long and Sykes suggest that the active sites responsible for the dissociation steam chemisorption, which is described by step 15 (Reaction 28), are carbon atoms that can readily provide a bond approaching the order of unity. These investigators noted that the opposite is true for the reaction of carbon dioxide with carbon, which occurs by the oxygen-exchange reaction in step 5 (Reaction 18) on active sites capable of forming bonds of an order greater than unity. This result has been postulated to explain why some investigators have observed that carbon monoxide and hydrogen inhibit the C–CO_2 reaction, whereas only hydrogen inhibits the carbon–steam reaction, at least with coconut charcoal. Other investigators using coal chars, however, have found that carbon monoxide significantly inhibited the carbon–steam reaction by carbon monoxide [7, 201], which may indicate that both the nature of the active sites and the reaction mechanism of importance depend on the nature of the carbon used.

The mechanism most widely proposed to describe the gasification of carbon in gases containing steam and hydrogen assumes the following sequence [188, 203, 204]:

Step 12. $\qquad H_2O + C_f \underset{k_{12}'}{\overset{k_{12}}{\rightleftarrows}} H_2 + C(O) \qquad$ [30]

Step 6. $\qquad C(O) \underset{k_6'}{\overset{k_6}{\rightleftarrows}} CO + C_f \qquad$ [19]

Step 12 (Reaction 30) is analogous to the oxygen–exchange reaction in step 5) (Reaction 18) in CO_2 and CO, and step 6 (Reaction 19) is the same as that assumed to occur in the C–CO_2 reaction. In the presence of steam and hydrogen alone, the preceding sequence leads to the following rate expression:

$$W = \frac{k_{12} n_t P_{H_2O}}{1 + (k_{12}'/k_6) P_{H_2} + (k_{12}/k_6) P_{H_2O}} \qquad [145]$$

142 FUNDAMENTALS OF COAL GASIFICATION

By comparison with Equation 137,

$$k_{OH} = k_{12} n_t \qquad [146]$$

$$K_{OH,1} = \frac{k'_{12}}{k_6} \qquad [147]$$

$$k_{OH,2} = \frac{k_{12}}{k_6} \qquad [148]$$

When carbon monoxide and carbon dioxide are also present in gas mixtures (in addition to steam and hydrogen), step 5 (Reaction 18) should also be considered:

Step 5. $\qquad CO_2 + C_f \underset{k'_s}{\overset{k_s}{\rightleftarrows}} CO + C(O) \qquad [18]$

which leads to the total gasification rate expression:

$$W = \frac{k_{12} n_t P_{H_2O}[1 - (k'_6 k'_{12}/k_6 k_{12})(P_{H_2} P_{CO}/P_{H_2O})] + k_5 n_t P_{CO_2}[1 - (k'_5 k'_6/k_5 k_6)(P^2_{CO}/P_{CO_2})]}{1 + (k_5/k_6)P_{CO_2} + [(k'_5/k_6) + (k'_6/k_6)]P_{CO} + (k'_{12}/k_6)P_{H_2} + (k_{12}/k_6)P_{H_2O}}$$
$$[149]$$

It can be seen that Equation 149 reduces to Equation 145 if only steam and hydrogen are present, and to Equation 112 if only carbon dioxide and carbon monoxide are present. Thus, in general, Equation 149 represents the sum of gasification rates by the carbon–steam (C–H_2O) reactions (first term) and the C–CO_2 reaction (second term).

Ergun [188] used the model containing reaction steps 5, 6, and 12 (Reactions 18, 19, and 30) to analyze kinetic results obtained for the gasification of metallurgical coke in a continuous fixed-bed integral gas conversion system, using various feed gases, at atmospheric pressure. With CO_2 as the feed gas, it was assumed that the following reduced form of Equation 149 applies under nonequilibrium conditions for the C–CO_2 reaction:

$$W = \frac{k_6 n_t}{1 + (k'_5/k_5)(P_{CO}/P_{CO_2})} \qquad [150]$$

Using this expression is equivalent to assuming that step 5 (Reaction 18) is in equilibrium and that step 6 (Reaction 19) is rate limiting. For feed gases containing steam, it was also assumed that step 12 (Reaction 30) is in equilibrium.

KINETICS OF GASIFICATION REACTIONS   143

The assumption of equilibrium in both oxygen-exchange reaction steps 5 and 12 leads to the following result:

$$\frac{k'_5 P_{CO}}{k'_5 P_{CO_2}} = \frac{k'_{12} P_{H_2}}{k_{12} P_{H_2O}} \qquad [151]$$

or

$$\frac{P_{CO_2} P_{H_2}}{P_{CO} P_{H_2O}} = \frac{k'_5 k_{12}}{k_5 k'_{12}} = K_{5,12} \qquad [152]$$

This implies that the water-gas shift equilibrium should be maintained in the gas phase. With these assumptions it follows that the rate of gasification in gases containing CO, CO_2, H_2, and H_2O can also be described by the following expressions for the $C-H_2O$ reaction and the $C-CO_2$ reaction under nonequilibrium conditions:

$$W = \frac{k_6 n_t}{1 + (k'_5 P_{CO} + k'_{12} P_{H_2}/k_5 P_{CO_2} + k_{12} P_{H_2O})} \qquad [153]$$

$$W = \frac{k_6 n_t}{1 + (k'_{12} P_{H_2}/k_{12} P_{H_2O})} \qquad [154]$$

$$W = \frac{k_6 n_t}{1 + (k'_5 P_{CO}/k_5 P_{CO_2})} \qquad [150]$$

To analyze results obtained in an experimental system where there was significant conversion of the feed gas, Ergun employed an integration form of the differential rate expression, based on an assumed gas-solid contacting mode, to derive values for the fundamental kinetic parameters. Ergun's values for the oxygen-exchange equilibrium constant, k_{12}/k'_{12}, are given in Table 23. He found these values to be reasonably consistent with values of the equilibrium constant obtained from gasification in CO/CO_2 mixtures, k_5/k'_5, by assuming the following relationship:

$$\frac{k_5}{k'_5} = \frac{k_{12}/k'_{12}}{K_{5,12}} \qquad [155]$$

where $K_{5,12}$ = equilibrium constant for the water-gas shift reaction.

Values of $k_6 n_t$ determined in experiments that used only steam as the feed gas, based on the assumption that Equation 154 is applicable, were found to be about 60% higher than values of $k_6 n_t$ determined with only CO_2 in the feed

TABLE 23 COMPARISON OF EXPERIMENTAL EQUILIBRIUM CONSTANTS FOR OXYGEN EXCHANGE IN CO_2-C REACTION WITH VALUES CALCULATED FROM H_2O-C REACTION[a]

Temperature (°C)	H_2O–C Reaction			CO_2–C Reaction
	k_{12}/k'_{12}	$K_{5,12}$	$k_5/k'_5 = [(k_{12}/k'_{12})/K_{5,12}]$	k_5/k'_5
1000	0.23	0.58	0.40	0.54
1100	0.39	0.47	0.83	1.0
1200	0.80	0.39	2.0	1.8

[a]Data from Menster and Ergun [177] and Ergun [188].

gas, using Equation 150. The relative effect of temperature on values of $k_6 n_t$, determined with either CO_2 or H_2O in the feed gas was quite similar however, corresponding to an activation energy of approximately 31 kcal/g-mol at 1000-1200°C.

The higher apparent values of $k_6 n_t$ obtained with steam as the feed gas compared to carbon dioxide as the feed gas were interpreted by Ergun as possibly being due to: (1) a polymodal distribution of active sites, with steam capable of reacting with more types than carbon dioxide or (2) various types of diffusion inhibition effects. Ergun also obtained results with various feed-gas mixtures of steam and CO, CO_2, or H_2. Using the kinetic parameters obtained when only steam was present in the feed gas, it is possible to compare predicted integral gasification rates with rates obtained experimentally with the gas mixtures. In general, the predicted rates are greater than the experimental rates with both CO_2-H_2O and CO-H_2O feed-gas mixtures and less than experimental rates with H_2-H_2O mixtures. It was not determined whether these differences are due to the use of an incorrect reaction mechanism, to the assumptions made in developing the rate expressions, or to other factors such as diffusion and/or gas-solid contacting phenomena or a distribution of active site types. These results illustrate the complications of analyzing and correlating data obtained in heterogeneous systems with no fixed stoichiometry.

At elevated pressures, the gasification of carbon with gases containing steam becomes even more complicated, because of the formation of methane. Blackwood and McCrory [200] suggested that reaction steps 1, 2, and 17 (Reactions 12, 13, and 31) occur in addition to steps 5, 6, and 12 (Reactions 18, 19, and 30) (to account for the synergistic interaction of steam and chemisorbed hydrogen to produce methane:

Step 17. $\qquad H_2O + C(H_2) \underset{k'_{17}}{\overset{k_{17}}{\rightleftarrows}} CH_4 + C(O) \qquad$ [31]

KINETICS OF GASIFICATION REACTIONS   145

Step 1.
$$H_2 + C_f \underset{k_1'}{\overset{k_1}{\rightleftharpoons}} C(H_2) \quad [12]$$

Step 2.
$$C(H_2) + H_2 \underset{k_2'}{\overset{k_2}{\rightleftharpoons}} CH_4 + C_f \quad [13]$$

Given this, the overall gasification rates can be described by the following complex expressions:

$$W = W_{CO_x} + W_{CH_4} \quad [156]$$

$$W_{CH_4} = \frac{\begin{aligned}&\{(k_6 + k_5'P_{CO} + k_{12}'P_{H_2} + k_{17}P_{CH_4})(k_1 k_2 P_{H_2}^2 - k_1' k_2' P_{CH_4})n_t \\ &+ k_1 k_{17} P_{H_2} P_{H_2O}(k_5' P_{CO} + k_6 + k_{12}' P_{H_2})n_t \\ &- k_1' k_{17}' P_{CH_4}(k_5 P_{CO_2} + k_6' P_{CO} + k_{12} P_{H_2O})n_t\}\end{aligned}}{\varphi} \quad [157]$$

$$W_{CO_x} = \frac{\begin{aligned}&\{(k_1' P_{H_2} + k_2 P_{H_2} + k_{17} P_{H_2O}) \cdot [k_6(k_{12} P_{H_2O} + k_5 P_{CO_2}) \\ &- k_6' P_{CO}(k_{12}' P_{H_2} + k_5' P_{CO})] n_t + k_6 k_{17} P_{H_2O}(k_1 P_{H_2} \\ &+ k_2' P_{CH_4}) n_t - k_6' k_{17}' P_{CO} P_{CH_4}(k' + k_2 P_{H_2})n_t\}\end{aligned}}{\varphi} \quad [158]$$

$$\varphi = (k_1' + k_2 P_{H_2}) \cdot (k_5 P_{CO_2} + k_5' P_{CO} + k_6 + k_6' P_{CO} + k_{12} P_{H_2O} + k_{12}' P_{H_2})$$
$$+ (k_1 P_{H_2} + k_2' P_{CH_4})(k_5' P_{CO} + k_6 + k_{12}' P_{H_2}) + (k_{17} P_{H_2O})$$
$$\cdot (k_1 P_{H_2} + k_2' P_{CH_4} + k_5 P_{CO_2} + k_5' P_{CO} + k_6 + k_6' P_{CO} + k_{12} P_{H_2O}$$
$$+ k_{12}' P_{H_2}) + (k_{17}' P_{CH_4})(k_1 P_{H_2} + k_1' + k_2 P_{H_2} + k_2' P_{CH_4} + k_5 P_{CO_2}$$
$$+ k_6' P_{CO} + k_{12} P_{H_2O}) \quad [159]$$

where

W_{CH_4} = net gasification rate leading to methane formation, in mol carbon/(mol carbon-min)

W_{CO_x} = net gasification rate leading to (CO + CO_2 formation, in mol carbon/(mol carbon-min)

The terms in the numerator of Equation 157 suggest that carbon gasification leading to net methane formation occurs by four paths equivalent to the following overall reactions:

Steps 1, 2.
$$2H_2 + C \longrightarrow CH_4 \quad [12, 13]$$

146   FUNDAMENTALS OF COAL GASIFICATION

Steps 1, 12, 17.	$2H_2 + C \longrightarrow CH_4$	[12, 30, 31]
Steps 1, 6, 17.	$2C + H_2 + H_2O \longrightarrow CO + CH_4$	[12, 19, 31]
Steps 1, 5, 17.	$C + H_2 + H_2O + CO \longrightarrow CH_4 + CO_2$	[12, 18, 31]

Similarly, Equation 151 suggests that carbon gasification leading to net carbon oxide formation occurs by four paths equivalent to these overall reactions:

Steps 6, 12.	$H_2O + C \longrightarrow CO + H_2$	[19, 30]
Steps 2, 6, 17.	$H_2O + C \longrightarrow CO + H_2$	[13, 19, 31]
Steps 5, 6.	$CO_2 + C \longrightarrow 2CO$	[18, 19]
Steps 1, 6, 17.	$C + H_2 + H_2O \longrightarrow CO + CH_4$	[12, 19, 31]

Equations 157 and 158 also reduce to certain rate expressions proposed for the simpler reaction systems. For example, in the presence of CO and CO_2 alone, Equation 158 reduces to Equation 112; in the presence of H_2 and CH_4 alone, Equation 157 reduces to Equation 81; in the presence of only CO, CO_2, H_2, and H_2O, and neglecting reaction steps 1, 2, and 17 (Reactions 12, 13, and 31) reduces to Equation 149.

Equations 157 and 158 define the net rates of carbon gasification to form CH_4 and $(CO + CO_2)$, respectively. In order to define the net formation rates of all gaseous species, due only to reaction steps 1, 2, 5, 6, 12, and 17 (Reactions 12, 13, 18, 19, 30, and 31), it is necessary to derive an additional expression, depending on the model considered. An example of such an expression which defines the net rate of CO_2 formation, is given as follows:

$$W_{CO_2} = \frac{(k_1' + k_2 P_{H_2} + k_{17} P_{H_2O})[k_5' P_{CO}(k_6' P_{CO} + k_{12} P_{H_2O}) - k_5 P_{CO_2}(k_6 + k_{12}' P_{H_2})] n_t + (k_1' + k_2 P_{H_2})[k_5' k_{17} P_{CO} P_{H_2O} - k_5 k_{17}' P_{CO_2} P_{CH_4}] n_t}{\varphi} \quad [160]$$

The net rates of formation of CO, H_2, and H_2O can be defined from Equations 157, 158, and 160, respectively

$$W_{CO} = W_{CO_x} - W_{CO_2} \quad [161]$$

$$W_{H_2} = W_{CO_x} + W_{CO_2} - W_{CH_4} + \lambda(W_{CO_x} + W_{CH_4}) \quad [162]$$

$$W_{H_2O} = -W_{CO_x} - W_{CO_2} \quad [163]$$

where λ is the ratio of hydrogen in char to carbon in char (in mol/mol).

The preceding expression given for the differential formation of gaseous species would not, of course, individually apply under conditions where homogeneous reactions occur in the gas phase, or where catalytic reactions occur possibly influenced by char mineral matter or reactor walls. The water-gas shift reaction (Reaction 3) is an important example of a reaction that could be catalyzed in coal-char-gasification systems. Although this reaction has often been found to be approximately in equilibrium in gasification systems at temperatures above around 800°C, equilibrium cannot always be assumed, particularly in fluidized beds and under conditions where relatively little steam decomposition occurs [3]. At present, however, there is no general way to reliably estimate water-gas shift equilibrium approach ratios as a function of the many kinetic and mass-transfer parameters that may affect such ratios. Although simplified empirical correlations have been proposed to define equilibrium approach factors for this reaction in char gasification systems [3], such correlations cannot be considered generally applicable and should, therefore, be used with caution.

It has not been determined whether Equations 157, 158, and 160 are applicable under conditions where all the major gas species are present, particularly at elevated pressures, where carbon gasification to form methane becomes significant. This is, of course, understandable in view of the large number of kinetic parameters involved, which tends to dilute the fundamental significance of numerical evaluations made of such parameters. It is also apparent that even a relatively slight alteration of the mechanism described can significantly alter the form of the resulting rate expressions, particularly if more than one type of active site is postulated or if nonuniform surface properties are important. Even at low pressures, the consideration of nonuniform surfaces with respect to reaction energetics can lead to unusual rate expressions, such as that given by Equation 138 [205, 206]. Because of such complications, kinetic data obtained for the gasification of carbons, under conditions where both methane and carbon oxides are formed simultaneously [3, 7, 174, 200, 201, 207-212] have been correlated in some studies by what was intended, or probably should be considered, as empirical expressions. These give relatively little direct insight into the reaction mechanisms.

In an extensive study conducted at the Consolidation Coal Company [174, 207, 208, 210], the gasification in steam-hydrogen mixtures of a Disco char derived from a Pittsburgh seam coal was investigated in a laboratory fluidized bed at 816-927°C and 1-30 atm. Differential gasification rates determined by an extrapolation procedure were reported by Zielke and Gorin [174] in terms of the following empirical expressions:

$$W_T \text{ (total carbon gasification rate)} = AP^n \qquad [164]$$

$$W_{CH_4} \text{ (methane formation rate)} = DP^m \qquad [165]$$

$$W_{CO+CO_2} \text{ (carbon oxide formation rate)} = W_T - W_{CH_4} \qquad [166]$$

where P is the total pressure and A, D, m, and n are functions of total carbon conversion and gas composition at a given temperature but are independent of total pressure. In this study, values for the constants A, D, m, and n were represented graphically.

In another study by Johnson [7], empirical kinetic correlations were developed to describe differential coal-char gasification rates, based on results obtained from several hundred tests in a thermobalance apparatus. Numerical evaluations of the kinetic parameters in the proposed correlations were determined from results obtained with a variety of bituminous coal chars gasified in H_2-CH_4 mixtures, H_2-H_2O mixtures, and CO-CO_2-H_2-H_2O-CH_4 mixtures, at temperatures of 800-1100°C and total pressures of from 1-70 atm. The formulation of this model with respect to gasification in hydrogen–methane mixtures was discussed previously.

The model assumes that in gas mixtures containing hydrogen and steam, three reversible gasification reactions occur independently of one another:

$$H_2O + C \rightleftharpoons CO + H_2 \qquad [32]$$

$$2H_2 + C \rightleftharpoons CH_4 \qquad [33]$$

$$H_2 + H_2O + 2C \rightleftharpoons CO + CH_4 \qquad [34]$$

It is further assumed in this model that the gasification of carbon by carbon dioxide is negligible under the conditions considered and that carbon dioxide interacts only in the water-gas shift reaction.

Total carbon conversion rate, dx/dt, is described by the following expression:

$$\frac{dx}{dt} = f_L k_T (1 - X)^{2/3} e^{-\alpha x^2} \qquad [167]$$

where

f_L = relative reactivity factor, which depends on the char type and char pretreatment temperature

α, k_t = kinetic parameters

Also

$$k_T = k_{(32)} + k_{(33)} + k_{(34)} \qquad [168]$$

where $k_{(32)}$, $k_{(33)}$, and $k_{(34)}$ are the rate constants for the three gasification reactions under consideration (Reactions 32-34). Thus knowledge of $k_{(32)}$, $k_{(33)}$, and $k_{(34)}$ not only gives the total gasification rate, but also defines reaction stoichiometry.

KINETICS OF GASIFICATION REACTIONS

Although the preceding correlations indicate that Reactions 32–34 occur independently, implying that different active sites are involved, the rate of each reaction is assumed to be proportional to the same carbon conversion term, $(1-X)^{2/3}\exp(-\alpha X^2)$. Individual parameters in Equations 167 and 168 are defined as a function of the temperature and partial pressure of individual gas species, as follows ($k_{(32)}$ in min^{-1}):

$$f_L = f_L^0 \exp\left[-\frac{9340}{R}\left(\frac{1}{T} - \frac{1}{T_P}\right)\right] \qquad [169]$$

$$k_{(32)} = \frac{\exp(9.0201 - 12{,}910/T)[1 - (P_{CO}P_{H_2}/P_{H_2O}K_{(32)}^E)]}{\{1 + \exp(-22.216 + 24{,}882/T)[(1/P_{H_2O}) + 16.35(P_{H_2}/P_{H_2O}) + 43.5(P_{CO}/P_{H_2O})]\}^2} \qquad [170]$$

$$k_{(33)} = \frac{P_{H_2}^2 \exp(2.6741 - 13{,}672/T)[1 - (P_{CH_4}/P_{H_2}^2 K_{(33)}^E)]}{1 + P_{H_2}\exp(-10.452 + 11{,}098/T)} \qquad [171]$$

$$k_{(34)} = \frac{P_{H_2}^{1/2}P_{H_2O}\exp(12.4463 - 20{,}043/T)[1 - (P_{CH_4}P_{CO}/P_{H_2}P_{H_2O}K_{(34)}^E)]}{1 + \exp(-6.6676 + 8{,}443/T)[P_{H_2}^{1/2} + 0.85P_{CO} + 18.62(P_{CH_4}/P_{H_2})]^2} \qquad [172]$$

$$\alpha = \frac{52.7P_{H_2}}{1 + 54.3P_{H_2}} + \frac{0.521P_{H_2}^{1/2}P_{H_2O}}{1 + 0.707P_{H_2O} + 0.5P_{H_2}^{1/2}P_{H_2O}} \qquad [173]$$

where

$K_{(32)}^E, K_{(33)}^E, K_{(34)}^E$ = equilibrium constants for Reactions 32–34 considering carbon as graphite

T = reaction temperature (in °K)

T_p = maximum temperature to which the char has been exposed prior to gasification in °K (if $T_p < T$, then a value of $T_p = T$ is used)

P_i = partial pressure of the gas-phase species in atmospheres

f_L^0 = relative reactivity factor, which depends on the nature of the particular carbonaceous solid

t = time (in min)

As can be seen from the above description, Johnson has proposed that the effect of pretreatment temperature on the rates of gasification in gases containing

150 FUNDAMENTALS OF COAL GASIFICATION

steam and hydrogen is the same as that proposed for gasification in hydrogen alone. Although fewer different types of coal chars were tested in gases containing steam and hydrogen than were tested in hydrogen alone, Johnson nevertheless suggested that the relative-reactivity-factor correlation (Figure 50) for gasification in hydrogen also appears to be applicable for gasification in steam-hydrogen mixtures.

In addition to the thermobalance data used to develop the model described, Johnson also considered the differential rate data of Zielke and Gorin [174] for the gasification of Disco char. Zielke and Gorin's data exhibited the constant or decreasing specific gasification rates with increasing carbon conversion that are predicted by Equations 167 and 173 as a function of gas composition and pressure. A comparison of the predictions of Johnson's correlation with the initial gasification rates ($X = 0$) reported by Zielke and Gorin is shown in Figures 61 and 62 for the gasification of Disco char at 917°C. The value of f_L^0 was equal to

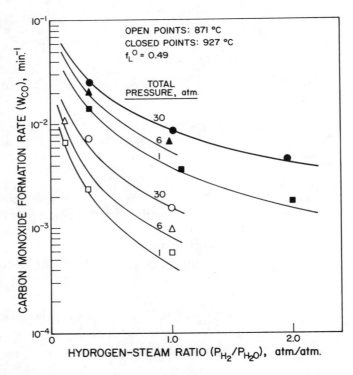

Figure 61 Comparison of experimental and calculated carbon monoxide formation rates for gasification of Disco char in steam–hydrogen mixtures. Data from Zielke and Gorin [174] and Johnson [7].

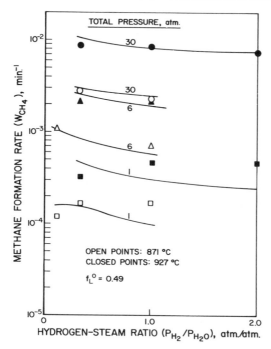

Figure 62 Comparison of experimental and calculated methane formation rates for gasification of Disco char in steam–hydrogen mixtures. Data from Zielke and Gorin [174] and Johnson [7].

0.49. Although Zielke and Gorin's complete data show that relative values of $W_{CH_4} = W_{CO}$ increase somewhat with increasing carbon conversion, contrary to Johnson's model (which predicts that this ratio should not vary under constant environmental conditions), it has been suggested by investigators at the Consolidation Coal Company [213] that this effect may be due to a catalytic reaction downstream of the fluidized-bed reactor, in which some of the carbon monoxide produced in the reactor was converted to methane.

For gasification in pure steam, Johnson's correlation predicts that rates initially increase with pressure but eventually become independent of pressure at pressures greater than 10 atm, as shown in Figure 63 ($f_L^0 = 1$). Similar behavior was observed by Van Heek, Jüntgen, et al. [209] for the gasification of various coals or coal chars in steam, under conditions of constant coal heatup rate. Figure 64 compares the initial specific gasification rate in pure steam at 10 atm as predicted by Johnson's [7] correlation ($f_L^0 = 1$) with that predicted by Van Heek's correlation [209] for the gasification of a high-volatility bituminous coal.

152 FUNDAMENTALS OF COAL GASIFICATION

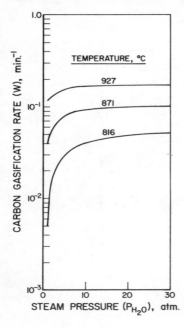

Figure 63 Effect of pressure on carbon gasification rate in pure steam $(f_L 0 = 1.0)$. Data from Johnson [7].

The kinetics of the gasification of coconut charcoal in steam–hydrogen mixtures were studied by Blackwood [211] and by Blackwood and McGrory [200] at total pressures of 1–50 atm and at temperatures of 750–850°C. Of the carbon gasified in this study at the relatively low hydrogen: steam ratios employed (below about 0.1 mol/mol), less than approximately 20% was converted to methane, with the remaining gas composed of carbon monoxide and carbon dioxide. Blackwood and McGrory suggest that carbon dioxide is a secondary product of the water–gas shift reaction. Relatively large gas flow rates were used in that study, to limit steam decomposition to less than about 6%, and results were interpreted as if the system were a differential reactor. Although total carbon conversion was usually less than about 10%, in some tests specific carbon conversion rates at 830°C were approximately constant for carbon conversions up to 50%.

That study was primarily interested in quantitative evaluations of the kinetics of carbon oxide formation. At constant steam partial pressure, carbon oxide formation rates decreased significantly with increasing hydrogen partial pressure, although this inhibition by hydrogen was less severe at 830°C than at 750°C. At constant hydrogen partial pressure, rates increased continuously with increasing steam partial pressure. At 830°C and a low hydrogen partial pressure, rates were essentially proportional to the steam partial pressure. At lower tem-

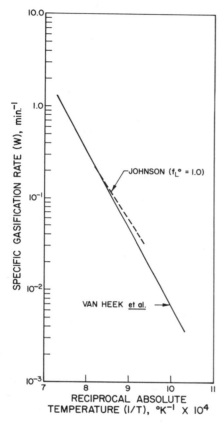

Figure 64 Comparison of gasification rates in pure steam at 10 atm. Data from Johnson [7] and Van Heek, Jüntgen, et al. [209].

peratures and higher hydrogen partial pressures, the reaction order with respect to steam partial pressure was greater than unity. Methane formation rates at 830°C were found to be relatively independent of the hydrogen partial pressure over the range employed but increased with steam partial pressure at a reaction order less than unity.

Blackwood and McGrory [200] postulated the following sequence to develop kinetic expressions for analyzing their results:

Step 15. $\quad C_f \text{ (dual site)} + H_2O \underset{k'_{15}}{\overset{k_{15}}{\rightleftharpoons}} C(H)(OH)$ [28]

Step 16. $\quad C(H)(OH) \overset{k_{16}}{\longrightarrow} C(H_2)(O)$ [29]

Step 6. $\quad C(O) \overset{k_6}{\longrightarrow} CO$ [19]

Step 1. $$H_2 + C_f \underset{k_2}{\overset{k_1}{\rightleftarrows}} C(H_2) \qquad [12]$$

Step 17. $$C(H_2) + H_2O \xrightarrow{k_{17}} CH_4 + C(O) \qquad [31]$$

Reaction steps 15 and 16 (Reactions 28 and 29) are the same as those proposed by Long and Sykes [197] to account for steam adsorption and decomposition on dual sites present on the carbon surface. In this model, carbon monoxide is formed by means of step 6 (Reaction 19), corresponding to the decomposition of the surface carbon–oxygen complex. Methane formation occurs by step 17 (Reaction 31), postulated to account for the interaction between surface carbon–hydrogen complex and gaseous steam to produce methane and surface oxygen–carbon complex. Although Blackwood and McGrory indicated that methane could also be formed by reaction step 2 (Reaction 13), they felt that this reaction would only be important at higher hydrogen partial pressures than those employed in their study.

Based on these reaction steps, the following expression was derived to describe the rate of carbon gasification to form carbon monoxide:

$$W_{CO} = \frac{\{k_{15}k_{16}/(k'_{15}k_{16})P_{H_2O} + (k_1 k_{17}/k'_1)P_{H_2}P_{H_2O} + (2k_{15}k_{16}k_{17}/k_5(k_{16} + k'_{15})P^2_{H_2O}\}n_t}{1 + (k_1/k'_1)P_{H_2} + \{(k_{17}/k_5) + [(k_{15}k_{16}/k_6) + (k_{15}k_{16}/k_5) + k_{15}/k'_{15} + k_{16}]\}P_{H_2O} + \{[2k_{15}k_{16}k_{17}/k_5 k_6(k'_{15} + k_{16})] + [k_{15}k_{17}/k_5(k_{15} + k_6)]\}P^2_{H_2O} + (k_1 k_{17}/k_5 k_6)P_{H_2}P_{H_2O}\}} \qquad [174]$$

The P_{H_2}, P_{H_2O}, and $P^2_{H_2O}$ terms in the denominator of Equation 174 were considered to be negligible, leading to the following equation form:

$$W_{CO} = \frac{k_{OH}P_{H_2O} + k'_{OH}P_{H_2}P_{H_2O} + k''_{OH}P^2_{H_2O}}{1 + K_{OH,1}P_{H_2} + K_{OH,2}P_{H_2O}} \qquad [175]$$

Although Blackwood and McGrory did not give an expression to describe methane formation rates, the following equation can be derived from their model:

$$W_{CH_4} = \frac{k'_{OH}P_{H_2}P_{H_2O} + (k''_{OH/2})P^2_{H_2O}}{1 + K_{OH,1}P_{H_2} + K_{OH,2}P_{H_2O}} \qquad [176]$$

where the kinetic parameters are defined in the same way as those in Equation 175. Blackwood and McGrory evaluated the kinetic parameters in their model by analyzing carbon monoxide formation rates and obtained the results given

KINETICS OF GASIFICATION REACTIONS

TABLE 24 VALUES OF KINETIC PARAMETERS

Temperature (°C)	k_{OH} (min^{-1} atm^{-1}) × 10^{-3})	k'_{OH} (min^{-1} atm^{-2}) × 10^{-3})	k''_{OH} (min^{-1} atm^{-2}) × 10^{-5})	$K_{OH,1}$ (atm^{-1})	$K_{OH,2}$ (atm^{-1})
750	0.43	0.36	1.80	35	0.06
790	1.50	0.60	1.80	35	0.09
830	4.44	1.26	1.80	35	0.14

in Table 24. The apparent activation energies for k_{OH}, k'_{OH}, and K^2_{OH} are 60, 30, and 20 kcal/g-mol, respectively.

Experimentally obtained gasification rates [7] at 830°C are compared to values calculated from Blackwood and McGrory's model [200] in Figures 65 and 66. In Figure 65 the predicted carbon monoxide formation rates are seen to be in good agreement with experimental values, which is to be expected since these data were used for the evaluation of kinetic parameters. Figure 66 shows

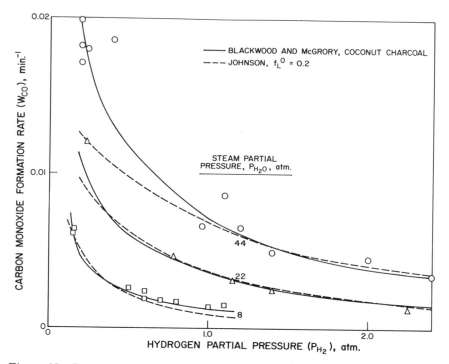

Figure 65 Comparison of carbon monoxide formation rates from gasification of carbons, based on different kinetic correlations. Data from Blackwood and McGrory [200] and Johnson [7].

156 FUNDAMENTALS OF COAL GASIFICATION

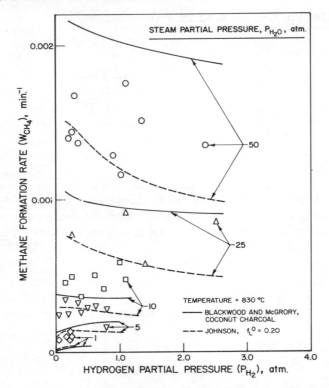

Figure 66 Comparison of methane formation rates from gasification of carbons, based on different kinetic correlations. Data from Blackwood and McGrory [200] and Johnson [7].

less consistency between experimentally determined methane formation rates and values predicted from Equation 176 using the kinetic parameters in Table 24, which were deduced from carbon monoxide formation data. The model does, however, predict the relative independence of methane formation rates from hydrogen partial pressure over the range studied. It is probable that the use of a different set of kinetic parameters derived both from carbon oxide and methane formation rates would show better consistency with all the data obtained.

Figures 65 and 66 also show predicted carbon oxide and methane formation rates based on Johnson's [7] model, using a value of $f_L^0 = 0.2$. In Figure 65 it can be seen that, except at high steam and low hydrogen partial pressures, a surprisingly good fit is obtained. Under the conditions given in Figure 66, the predictions of Johnson's model for methane formation rates are only approximate, as are those of Blackwood and McGrory's model. Johnson's model does, however, show that the predicted methane formation is relatively insensitive to variations in the hydrogen partial pressure over the range considered. The general similarity between the predictions of Blackwood and McGrory's model and Johnson's model (cf. Figures 65 and 66) is of particular interest in view of the significant differences in the forms of correlations used and in the numerical evaluation of kinetic parameters. This is a direct indication that the theoretical interpretation of kinetic parameters in correlations developed to describe carbon gasification in more complex systems should be made with caution.

Only a few studies conducted at elevated pressures have evaluated the effects of carbon monoxide and methane on the gasification of coal chars in gases containing steam and hydrogen. In Johnson's model, carbon monoxide severely inhibits the steam-carbon reaction, to an even greater extent than does hydrogen. This model also predicts that methane formation is inhibited by hydrogen, carbon monoxide, and methane itself, even at conditions far removed from equilibrium. Curran, Fink, et al. [201] have also studied effects of gas composition on the gasification of Renner Cove lignite char in gases containing steam, hydrogen, carbon monoxide, and methane at 816°C and at total pressures of 10-15 atm. The results of this study were interpreted by assuming that the following two reactions predominate:

$$C + H_2O \Longleftrightarrow CO + H_2 \quad [10]$$

$$2C + H_2 + H_2O \Longleftrightarrow CO + CH_4 \quad [25]$$

The rate-controlling step for Reaction 10 was assumed to be the reaction between adsorbed H_2O and a free active site on the carbon surface to produce an adsorbed carbon-oxygen complex and an adsorbed carbon-hydrogen complex. This model can be represented by the following sequence:

Step 13. $$C_f + H_2O \underset{k'_{13}}{\overset{k_{13}}{\rightleftarrows}} C(H_2O) \quad [26]$$

Step 1. $$C_f + H_2 \underset{k'_1}{\overset{k_1}{\rightleftarrows}} C(H_2) \quad [12]$$

Step 6. $$C(O) \underset{k'_6}{\overset{k_6}{\rightleftarrows}} CO + C_f \quad [19]$$

Step 20. $$C(H_2O) + C \underset{k'_{20}}{\overset{k_{20}}{\rightleftarrows}} C(O) + C(H_2) \quad [35]$$

Steps 1, 6, and 13 (Reactions 12, 19, and 26) were assumed to be rapid and thus in equilibrium.

The rate-controlling step for Reaction 25 was assumed to be the reaction between adsorbed hydrogen and adsorbed H_2O on adjacent sites to produce adsorbed carbon–oxygen complex and adsorbed methane. The model for this overall reaction can be represented by the sequence:

Step 13. $\qquad C_f + H_2O \underset{k'_{13}}{\overset{k_{13}}{\rightleftarrows}} C(H_2O) \qquad$ [26]

Step 1. $\qquad C_f + H_2 \underset{k'_1}{\overset{k_1}{\rightleftarrows}} C(H_2) \qquad$ [12]

Step 6. $\qquad C(O) \underset{k'_6}{\overset{k_6}{\rightleftarrows}} CO + C_f \qquad$ [19]

Step 19. $\qquad C(CH_4) \underset{k'_{19}}{\overset{k_{19}}{\rightleftarrows}} CH_4 + C_f \qquad$ [36]

Step 18. $\qquad C(H_2) + C(H_2O) \underset{k'_{18}}{\overset{k_{18}}{\rightleftarrows}} C(O) + C(CH_4) \qquad$ [37]

Steps 1, 6, 13, and 19 (Reactions 12, 19, 26 and 36) were also assumed to be rapid and in equilibrium.

In the models proposed it was assumed that Reactions 10 and 25 occur independently of one another on the carbon surface; that is, different active sites and surface complexes are involved in the two reactions sequences. With these assumptions, the following equations and evaluations of kinetic parameters were derived to describe initial gasification rates ($X = 0$):

$$W_C = \frac{0.0576[P_{H_2O} - (P_{CO}P_{H_2}/3.26)]}{[1 + 0.035 P_{H_2O} + 0.36 P_{H_2} + 1.30 P_{CO}]^2} \qquad [77]$$

$$W_{CH_4} = \frac{0.01564[P_{H_2O}P_{H_2} - (P_{CH_4}P_{CO}/0.388)]}{[1 + 0.19 P_{H_2O} + 0.16 P_{H_2} + 1.25 P_{CO} + 1.18 P_{CH_4}]^2} \qquad [78]$$

where

P_i = partial pressure (in atm) (i = CO, H_2, H_2O, CH_4)

W_C = rate of reaction given in Reaction 10 (in mol carbon gasified/mol carbon-min)

W = rate of methane formation via Reaction 25 (in mol carbon gasified/ mol carbon-min)

From the stoichiometry of Reactions 10 and 25, the net rate of carbon gasification to form carbon monoxide, W_{CO}, is given by

$$W_{CO} = W_C + W_{CH_4} \qquad [179]$$

and the total carbon gasification rate, W_T, by

$$W_T = W_C + 2W_{CH_4} \qquad [180]$$

The effect of carbon monoxide in strongly inhibiting the rate of reaction in Reaction 10 is similar to the predictions of Johnson's model for the same reac-

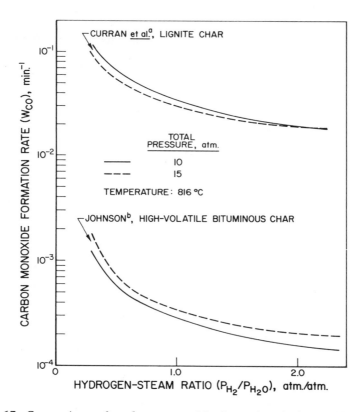

Figure 67 Comparison of carbon monoxide formation during gasification of different coal chars in steam-hydrogen mixtures. Data from Curran, Fink, et al. [201] and Johnson [7].

160 FUNDAMENTALS OF COAL GASIFICATION

tion. The correlation of Curran and colleagues also predicts that carbon monoxide and methane will strongly inhibit the methane formation reaction.

The numerical evaluations given in Equations 177 and 178 correspond only to initial gasification rates. Different sets of parameters were reported at higher carbon conversion levels, reflecting the fact that total specific gasification rates as well as relative methane and carbon monoxide formation rates were found to vary as a function of carbon conversion level. Such behavior, not usually observed with bituminous coal chars, may be related to variations in catalytic influences during the course of gasification of certain alkali-mineral constituents inherent to many lignites.

The general reactivity levels of the Renner Cove lignite char used by Curran and colleagues and of a typical char derived from a high-volatility bituminous coal can be compared on the basis of the correlations given in Equations 177 and 178 and the correlation proposed by Johnson ($f_L = 1$). Such a comparison is shown in Figures 67 and 68. Although the qualitative trends exhibited in Figure 67 for the two kinds of char are similar with respect to the gasification of carbon to form carbon monoxide, the lignite char is about 100 times more re-

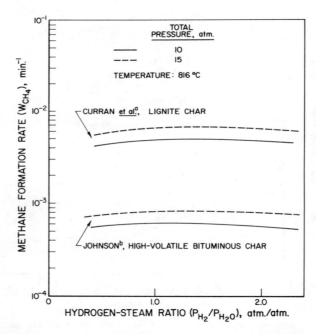

Figure 68 Comparison of methane formation rates during gasification of different coal chars in steam-hydrogen mixtures. Data from Curran, Fink, et al. [201] and Johnson [7].

active than the bituminous char. As will be discussed in a later section, this difference in reactivity level is due primarily to inherent catalytic influences on lignite-char gasification, influences that are not present to a significant extent in the gasification of most bituminous chars.

As shown in Figure 68, the reactivity of the lignite char with respect to methane formation is only about 8 times greater than that of the bituminous char. This reflects the selectivity of the catalyst present in lignite chars in favoring the formation of carbon monoxide and also points out that the kinetic correlations developed for bituminous-char gasification are not generally applicable to the gasification of lignite char.

Particle-size Effects on Char Gasification. The gasification of porous carbons involves the transport of reactant and product gases between the external bulk gas phase and the internal particle surfaces, where chemical reactions take place. Under conditions where gaseous concentrations within a particle differ from the external gaseous concentrations, mass-transport or diffusion processes affect overall gasification rates. When this occurs, overall gasification rates generally tend to decrease with increasing particle size. Gaseous concentration gradients can exist between either (1) the bulk gas phase and the external particle surface, due to film-diffusion resistances, or (2) the external particle surface and internal particle surfaces, due to diffusion limitations within the particle pore structures. In either case, where there are limitations on the gaseous diffusion in the gasification of carbon particles, this effect must be considered in interpreting results of laboratory studies and in applying chemical kinetic models to anticipate behavior in large-scale systems.

Film-diffusion effects can be reasonably well anticipated on the basis of existing technologies. The more complicated situation is in dealing with internal particle diffusion. Walker, Austin, et al [214] have described this situation in detail by considering diffusion effects on the gasification of purer carbons. Other investigators [174, 183, 215] have also quantitatively considered effects of intraparticle diffusion on the gasification of coal chars and other carbons, employing various theories [216–218] to account for gaseous mass transport within porous solids. With coal chars, however, the complex pore structures, which change during the course of gasification, present a formidable challenge to theoretical analysis of diffusion behavior.

Walker and co-workers [98, 99] have indicated that coal chars have a trimodal pore structure consisting of macropores; transitional, or "feeder" pores; and micropores, defined as being less than 12 Å in diameter. They have also suggested that, although gaseous diffusion through macropores and transitional (feeder) pores is quite rapid (a situation in which gaseous concentrations at the mouths of micropores closely approach the concentration at the external particle surface), gaseous diffusion within the micropore system can be quite slow.

In fact, for pores or apertures less than about 5 Å, gaseous diffusion is an activated process [219]. The micropore system within coal-char particles, however, contains almost all of the internal surface area and thus almost all of the active sites for gasification. As a general principle, then, it has been suggested that coal chars that contain large feeder pore systems tend to show less mass-transport limitations during gasification. The reason for this is that increasing the feeder pore channels decreases the average diffusion length within micropores. In this context it has been shown that chars derived from low-rank coals have larger feeder pore systems than chars derived from high-rank coals [98, 120].

The existence of diffusion limitations in the gasification of a carbon depends both on the chemical reactivity of the carbon and on the pore structure characteristic of the carbon. Thus in situations where chemical reactivity is low because of the inherent nature of the carbon or the reaction conditions employed, more severe diffusion resistance can be tolerated, without affecting gasification kinetics, than would be the case with high chemical reactivity. High chemical reactivity is, of course, favored by high temperatures. This leads to the observation that there is a transition in the controlling mechanism of gas-solid reactions from chemical-reaction control to diffusion control that occurs with increasing temperature. This transition is usually accompanied by a decrease in the overall temperature coefficient of reaction.

Because of the complex effects that occur in pore diffusion processes during coal-char gasification, quantitative predictions describing the interaction of chemical and diffusion phenomena require, at the very least, a detailed characterization of variations in the char pore structure. Although some information of this type is available, much more is needed. Fortunately, for many of the conditions relevant to coal-char-gasification systems that have been studied, relatively little effect of particle size on reaction kinetics has been observed. This is indirectly suggested by the large temperature coefficients derived to describe experimental results and the dependence of gasification rates on carbon conversion level, as well as by the more direct comparison of gasification rates obtained with different particle sizes. Van Heek, Jüntgen, et al. [209] for example, found that particle sizes of about 0.1-1.4 mm had no effect on the gasification of a bituminous coal in steam at 10-atm pressure and temperatures up to 950°C. The nominal gasification rate at 950°C was about 0.01 min^{-1}. Similarly, for a brown coal, particle size in the same range had no effect at temperatures up to 800°C, where a rate of 0.02 min^{-1} was obtained. Ergun [188] found that particle sizes of approximately 0.2-1 mm did not affect the gasification of a metallurgical coke in steam at 1 atm pressure and temperatures up to 1200°C. At 1200°C an integral carbon gasification rate of about 0.03 min^{-1} was obtained.

In spite of these results, there is little basis on which to accurately predict—even from an empirical standpoint—conditions under which internal particle diffusion may become important in coal-char-gasification systems. This is because there is only a limited number of studies in which the particle sizes of coal chars were systematically varied. Although very approximate considerations suggest that chemical-reaction control will predominate with most coal chars during gasification (uncatalyzed) at temperatures less than 1000°C and for particle sizes less than about 2 mm, the possibility of diffusion inhibition should be considered at more severe conditions, such as those likely to exist in steam-oxygen gasifiers.

Use of Catalysts in Coal-char Gasification. Existing and projected shortages of natural gas in the United States have stimulated extensive research to develop the technology for commercial production of synthetic high-Btu gases of pipeline quality. In this effort, coal gasification has been given particular emphasis because of the extensive availability of coal as a natural resource in this country, and several large-scale process-development programs are currently being conducted with government and industrial support. The main objective of these programs has been to use modern technologies to develop processes with significantly improved efficiencies, compared to classical European gasification systems. Data available from the different programs currently at various stages of development already indicate that the objectives of improved efficiency and performance will indeed be achieved in the relatively near future. This evidence also indicates, however, that there are still practical limitations to obtaining increased efficiency solely by manipulating process configurations and using modern gas-solid contacting systems, and that these limitations are due primarily to inherent chemistry of coal gasification in the absence of major catalytic influences. The fact that these limitations are not theoretical or themodynamic in nature but result from the relative and absolute chemical kinetic characteristics has created considerable interest in the use of catalysts in gasification systems as a means for further significant improvement in process efficiencies.

The primary process objective in the gasification of coal chars to produce methane is to achieve the following reaction stoichiometry:

$$2C + 2H_2O \rightleftharpoons CH_4 + CO_2 \qquad [38]$$

Unfortunately, this reaction is not feasible in uncatalyzed systems. The actual reaction path corresponds more closely to a three-part synthesis:

$$C + H_2O \rightleftharpoons CO + H_2 \qquad [10]$$
$$C + 2H_2 \rightleftharpoons CH_4 \qquad [1]$$
$$\underline{CO + H_2O \rightleftharpoons CO_2 + H_2 \qquad [3]}$$
$$2C + 2H_2O \rightleftharpoons CH_4 + CO_2 \qquad [38]$$

The reaction given by Reaction 10 is endothermic and does not normally occur with bituminous coals at reasonable rates below about 800°C. In fact, for most processes, temperatures of about 1000°C or higher are necessary to obtain the required conversions with a reasonable reactor size. This temperature level, however, does not favor the exothermic Reaction 1, which is relatively slower than Reaction 10 at elevated temperatures. Some processes improve this situation somewhat by using multizoned reactor systems, in which hydrogen generated in a high-temperature zone (by Reaction 10) subsequently reacts with carbon in a lower temperature zone to form methane (by Reaction 1). But even with this procedure, the overall gasification process is still endothermic. Additional heat to sustain the reactions—usually achieved by further combustion of the char with oxygen—requires increased feed coal and oxygen and increased catalytic methanation to produce a given amount of final methane product, all of which decrease the process thermal efficiencies.

The potential for approaching the stoichiometry indicated by Reaction 38 is possible only as a result of the action of catalysts that preferentially enhance either direct formation of methane, as in Reaction 1, or indirect formation of methane through a process such as Reaction 39, occurring in the gasifier itself:

$$CO + 3H_2 \rightleftharpoons CH_4 + H_2O \qquad [39]$$

Thermodynamic considerations, however, dictate that even with optimal catalytic action, substantial improvements in process efficiencies can occur only by operating the gasifier at temperatures much lower than those normally required in the absence of catalysts.

Since the 1920s various studies have investigated the effects of catalysts on the gasification of carbonaceous solids, with the objective of improving processes for the production of water gas, producer gas, or hydrogen as sources for producing ammonia. More recently, studies have also been conducted in connection with gas-cooled nuclear reactors [176, 220-232]. Although most work with these objectives has been conducted at low pressures with cokes, charcoals, and graphite, much of the information obtained—with the large number of catalysts evaluated—is of at least qualitative value in assessing the potential effects of catalysts on char gasification for methane production. Catalysts surveyed in the earlier work include metals, metal oxides, metal halides, alkali carbonates, and iron carbonyls, with particular emphasis on K_2CO_3, Na_2CO_3, KCl, NaCl, CaO, and transition metals. Some of the general conclusions are as follows:

1. Relative catalytic effects decrease with increasing gasification temperature.
2. In the gasification process, catalysts are generally more effective with gases containing steam than with gases containing hydrogen alone.

3. There is usually an intermediate optimum catalyst concentration, beyond which either negligible or negative effects result.
4. The relative effects of different catalysts can differ at different reaction conditions.
5. The specific methods and conditions used for catalyst impregnation can significantly affect subsequent gasification reactivity.
6. Catalyst impregnation is more effective than physical mixing with the carbon.

With respect to point (6), the method of catalyst addition is very important to the process of coal gasification. The advantages of catalyst impregnation must be economically weighed against the corresponding requirements for catalyst addition operations and catalyst recovery problems, which are likely to be more severe than with physically mixed catalyst–coal systems. The use of physical mixtures of catalysts and coal in laboratory studies presents a particular problem when the results are interpreted in terms of the performance of projected continuous gasifiers, because the transfer mechanism required for the catalyst to interact in the gasification step may depend on the physical nature of the contacting between the catalyst and the coal. Such contacting would be quite different for the fixed beds frequently used in laboratory studies than for the fluidized beds more likely to be used in large-scale systems.

Information that is directly relevant to the catalysis of coal-char gasification has been obtained only within the past few years. In one such investigation, conducted at Case Western Reserve University [233, 234], a thermobalance apparatus was used to study the effects of approximately 20 impregnated catalysts on coal-char-gasification kinetics. In these studies by Gardner, Samuels, et al. [233] and by Wilks, Samuels, et al [234] catalytic effects were investigated at 850–950°C and 500–1000 psi for gasification in either hydrogen or steam. Most tests used a 5% catalyst concentration. It was concluded that, although a few catalysts are as effective as the alkali metals in promoting gasification rates, none was found to be more active. With 5% $KHCO_3$, gasification rates in hydrogen at 500 psi and 950°C are increased twofold, whereas in steam, gasification rates are increased about threefold under the same conditions. More recently, data have been obtained indicating that catalyst combinations, such as alkaline earth metals with cobalt molybdate, are even more effective than single catalysts in enhancing gasification rates [235], suggesting that a study of catalyst combinations would be a promising area for future research.

A thermobalance apparatus was also used at the IGT to study the catalytic effects of exchangeable cations, such as sodium or calcium, on lignite-gasification reactivities [120]. This study was prompted by the fact that lignite chars are inherently more reactive than bituminous chars for gasification. However, when treated in acid (e.g., HCl), the gasification reactivity of the lignite chars is subtantially reduced and becomes comparable to the reactivity of the bituminous chars. It has been postulated that the inherently high reactivities of lignites are due to the catalytic influences of sodium or calcium combined with carboxyl functional groups in the orgainic structure of the lignites. When treated in acid, however, the cation can be removed by the following exchange reaction:

$$-COONa + H^+ \iff -COOH + Na^+ \qquad [40]$$

Since the concentration of carboxyl functional groups decreases significantly with increasing coal rank, this inherent catalytic effect is not significant in bituminous coal chars.

In studies conducted at the IGT the reversibility of this exchange reaction was used as a means to systematically investigate the effects of exchangeable cation concentration on lignite-char-gasification reactivities. Lignites were initially treated in hydrochloric acid to remove inherent calcium or sodium; controlled amounts of calcium or sodium were then added to the lignite by subsequent exchange, through reversal of Reaction 40. Experimental results showed that, for a 3% sodium concentration, a twentyfold increase in gasification rates was obtained at 760°C and 500 psi in a steam–hydrogen mixture, compared to rates obtained with the lignite when containing no exchangeable cations (acid-treated only).

In general, gasification rates increased with increasing catalyst concentration, at least up to a concentration of 3%. Similar results were obtained with both calcium and sodium at comparable mass concentrations for gasification in steam–hydrogen mixtures. Relative catalytic effects decreased with increasing gasification temperature, such that, with steam–hydrogen mixtures at 930°C, only a sevenfold increase in gasification rates was obtained with 3% catalyst addition. For gasification in pure hydrogen at 930°C, the catalyst had significantly less effect than with steam–hydrogen mixtures. Sodium was also found to be about 4 times more effective than calcium for gasification in pure hydrogen.

Attempts to add sodium and calcium to bituminous coals were unsuccessful, presumably because of the lack of carboxyl functional groups. It is not known whether the exceptionally large catalytic effects obtained with organically combined sodium and calcium in lignites are due to the particular chemical state of the catalytic complexes or to the greater dispersion that is likely, compared with simple deposition on internal surfaces by solution evaporation, but future studies to determine methods of promoting organic complexes between coal and catalyst appear to be promising.

Some recent studies have been conducted at several laboratories to investigate the gasification yields that can be obtained with physical catalyst-coal mixtures, under conditions of integral gas conversion at elevated pressures. In one such study conducted by the U.S. Bureau of Mines [236] the effects of physically mixed catalysts were investigated in a continuous, 4-in.-diameter, fluidized-bed reactor used in the development of the "synthane process." Additives included 2%, 5%, and 10% dolomitic limestone, 2% and 5% hydrated lime, and 5% lignite ash. For operation at 600 psi with a steam–oxygen feed at 875-950°C, total carbon conversions increased by an average of approximately 10%, with the added materials. This increase in conversion is equivalent to the effect that would be achieved with a decrease in gasifier temperatures of about 50°C, compared to the temperatures required for the same coal throughput without the additives. It was also noted that, with the majority of additives tested, no slag formation occurred, suggesting that the gasifier could be operated at higher temperatures, if desired, to achieve higher throughputs.

Another Bureau of Mines study, reported by Haynes, Gasior, et al. [237] evaluated the effects of approximately 40 different catalysts on yields obtained by the gasification of a high-volatility bituminous coal with steam at 850°C and 300 psig. In this study 5% catalyst concentrations were used in physical mixtures with the coal charge. A fixed-bed, flowing gas system was employed, and total gaseous yields were reported for 4-hr test periods. It should be noted that, since raw coals were used in this study as well as the one discussed earlier, total gas yields result from reactions that may occur in both the coal- and char-gasification stages.

As in earlier studies, Haynes and colleagues found that the best overall reactivities were achieved with alkali catalysts, particularly K_2CO_3 and KCl. With these materials, total carbon conversions of about 80% were obtained during the 4-hr test period, compared to a reference carbon conversion of less than 50% obtained without any catalyst. For the most part, increased activity led to increased formation of carbon oxides and hydrogen, relative to increases in methane formation.

An exception to this trend was obtained with an unactivated Raney nickel catalyst sprayed onto the surface of a metal tubular insert in the coal bed. In this case, total methane yields increased by about 24%, although total carbon conversion only increased by about 10%, indicating that the major influence of the catalyst was on the methanation of the carbon monoxide and hydrogen formed by direct gasification. The sprayed Raney nickel catalyst, however, decreased in reactivity with continued use, and loss of reactivity was estimated at 20-hr of operation. This loss of reactivity under the relatively severe conditions existing in the gasifier was attributed both to exfoliation of the catalyst and to sulfur poisoning, which is currently one of the major problems in the use of nickel-based catalysts in coal gasifiers.

In evaluating the significance of the role played by catalysts in coal gasification systems with gases containing both steam and hydrogen, it is important to distinguish between two major aspects: (1) the increase in total carbon gasification rate and (2) the relative increase in hydrocarbon production. This latter aspect is particularly important because it contributes to increasing the thermal efficiency of a gasification process and in many instances is probably an overriding factor in determining the practicality of using catalysts in coal gasifiers. Thus for the Bureau of Mines studies discussed previously, it is significant that the increased carbon conversion apparently corresponded to a decrease in thermal efficiency for the overall gasification step (although, with one exception, additives generally tended to increase the total carbon conversion rates).

The use of catalysts at lower temperatures has been the subject of several studies conducted at the University of Wyoming [238-240] with the objective of promoting the overall optimum reaction:

$$2C + 2H_2O \Longrightarrow CH_4 + CO_2 \qquad [38]$$

In one series of experiments, the gasification, with steam, of low-rank coals mixed with K_2CO_3 and commercial nickel catalyst was studied in a fixed-bed, flowing-gas system at 650°C. It was possible with this sytem, even at low pressure (30 psi), to produce a gas consisting primarily of CO_2 and CH_4, with optimum catalyst concentrations of about 1 g of nickel catalyst/g of coal and 0.2-0.25 g of K_2CO_3/g of coal. On a CO_2-free basis, product gases were consistently obtained with heating values of about 850 Btu/scf (British thermal units/standard cubic foot). At 650°C, however, these yields were obtained for total test periods of about 7.5 hr and for total carbon conversions of less than 50%. Such results suggest that, for operation at this low temperature, a much more active catalyst system or coal-catalyst contacting technique would be required for acceptable large-scale operation.

Two other major problems also inherent to this system are catalyst recovery and loss of catalyst activity. Under optimum reaction conditions, only about 80% of the potassium charged was recovered by a 1-hr water wash at room temperature, a relatively serious problem considering the large amounts of K_2CO_3 charged. In addition, relatively large fractions of the coal's sulfur combined with the nickel catalyst during gasification. Although this did not result in noticeable decreases in reactivity during batch charge tests, considering the large amounts of nickel catalysts employed, extrapolation of these effects indicated that replacement of the completely sulfided catalyst would be economically unfeasible, and that some means of catalyst regeneration would have to be employed in commercial application. Several approaches to dealing with catalyst poisoning by sulfur have not yet been completely satisfactory, and it has been suggested that perhaps a combination of the use of sulfur scavengers, catalyst regeneration, more powerful sulfur-resistant catalysts, and prior sulfur removal may be required as a final solution.

KINETICS OF GASIFICATION REACTIONS 169

The preceding observations are generally consistent with other recent evaluations [241-243] that conclude that there is a substantial need for additional research in areas related to the use of catalysts in coal-char gasification. Specifically, these needs include the development of new catalysts and catalyst systems, with particular emphasis on detailed investigations of the mechanisms involved in the interactions of catalysts with coal- or char-gasification processes. Such detailed considerations would ideally involve a much more fundamental characteristization of reaction systems than has been done in most of the previous studies. For example, particular attention should be directed toward characterization of the coals and coal chars used, in terms of not only elemental composition or rank type, but also petrographic composition and internal physical properties, such as surface area and pore volume distribution. The dispersion characteristics, chemical state, and physical properties of the catalysts employed should also be determined, not only in the raw coal, but also during the course of gasification.

With those catalyst systems having the desired operating characteristics, detailed kinetic studies would be needed to provide the information necessary to fully evaluate their application in an integrated commercial system over the range of environmental conditions to which the catalyst would typically be exposed. Such kinetic studies should include not only major reactions, but also side reactions and reactions pertinent to catalyst recovery.

REFERENCES

1. Krikorian, O. H., U.S. Atomic Energy Commission Report No. UCID-16587, National Technical Information Service, Springfield, Va., 1972.
2. Wen, C. Y., Abraham, O. C., and Talwalkar, A. T., *Am. Chem. Soc., Div. Fuel Chem., Prepr.,* **10**(4), 168-185 (1966); *Adv. Chem. Ser.,* No. 69 (American Chemical Soc., Washington, D.C.), 251-171 (1967).
3. Squires, A. M., *Transact. Inst. Chem. Eng. (Lond.),* **39**, 3-9 (1961).
4. Blackwood, J. D., and McCarthy, D. J., *Aust. J. Chem.,* **19**, 797-812 (1966).
5. Birch, T. J., Hall, K. R., and Urie, R. W., *J. Inst. Fuel,* **33**, 422-435 (1960).
6. Browning, L. C., and Emmett, P. H., *J. Am. Chem. Soc.,* **73**, 581-583 (1951).
7. Johnson, J. L., *Advances in Chemistry Series*, No. 131, American Chemical Society, Washington, D.C., 1974, pp. 145-178.
8. von Fredersdorff, C. G., and Elliott, M. A., and Lowry, H. H., Eds., *Chemistry of Coal Utilization*, Suppl. vol., Wiley, New York, 1963, pp. 892-1022.
9. Stull, D. R., and Prophet, H., Eds., *JANAF Thermochemical Tables*, 23.s, 2nd ed., U.S. Dept. Commerce, U.S. Government Printing Office, Washington, D.C., 1971, 1141 pp.

10. Baron, R. E., Porter, J. H., and Hammond, O. H., Jr., *Chemical Equilibria in Carbon-Hydrogen-Oxygen Systems*, MIT Press, Cambridge, Mass., 1976.
11. Agroskin, A. A., and Goncharov, E. I., *Coke Chem. USSR*, 11, 16-20 (1965).
12. Batchelor, J. D., Yavorsky, R. M., and Gorin, E., *J. Chem. Eng. Data*, 4, 241-246 (1959).
13. Binford, J. S., Jr., Strohmenger, J. M., and Hebert, T. H., *J. Phys. Chem.*, 71, 2404-2408 (1967).
14. Clendenin, J. D., Barclay, K. M., Donald, H. J., Gillmore, D. W., and Wright, C. C., *Transact. 7th Annual Anthracite Conference Lehigh Univ.*, Bethlehem, Pa., 1949, pp. 12-67.
15. Coles, G., *J. Soc. Chem. Ind.*, 42, 435-439 (1923).
16. Fritz, V. W., and Moser, H., *Feuerungstechnik*, 5, 97-107 (1940).
17. Gomez, M., Gayle, J. B., and Taylor, A. R., Jr., *U.S. Bur. Mines, Rept. Invest.* (6607), 1-45 (1965).
18. Kazmina, V. V., *Coke Chem. USSR*, 11, 26-29 (1965).
19. Mantell, C. L., *Industrial Carbon*, 2nd ed., Van Nostrand, New York, 1947, p. 432.
20. Porter, H. C., and Taylor, G. B., *Ind. Eng. Chem.*, 5, 289-293 (1913).
21. Terres, E., Dahne, H., Nandi, B., Scheidel, C., Trappa, K., and Rauth, P., *Brennstoff-Chem.*, 37, 366-370 (1956).
22. Terres, E., Dahne, H., Nandi, B., Scheidel, C., and Trappe, K., *Brennstoff-Chem.*, 37, 269-277 (1956).
23. Voloshin, A. K., Virozub, I. V., Kazmina, V. V., and Kurbatova, M. Y., *Coke Chem. USSR*, 3, 17-20 (1962).
24. Dewey, P. H., and Harper, D. R., III, *J. Res. Nat. Bur. Standards*, 21, 457-474 (1938).
25. Morlock, R. J., Naso, A. C., and Cameron, J. R., *Symp. Sci. Tech. Coal*, Canada Department of Energy, Mines, and Resources, Mines Branch, Ottawa, 1967, pp. 127-131.
26. Lee, A. L., *Am. Chem. Soc., Div. Fuel Chem., Prepr.*, 12(3), 19-31 (1968).
27. Mahajan, O. P., Tomita, A., and Walker, P. L., Jr., *Fuel*, 55, 63-69 (1975).
28. Kirov, N. Y., *Br. Coal Util. Res. Assoc., Mon. Bull.*, 29(2), 33-57 (1965).
29. Cragoe, C. S., *U.S. Bureau of Standards Miscellaneous Publication No. 97*, 1929.
30. Clendenin, J. D., Barclay, K. M., Donald, H. J., Gilmore, D. W., and Wright, C. C., *Penn. State Univ., Mineral Ind. Exp. Sta., Tech. Paper No. 160*, 11-67 (1949).
31. Institute of Gas Technology, *Preparation of a Coal Conversion Systems Technical Data Book, Final Report, October 31, 1974-April 30, 1976*, U.S. Energy Res. Develop. Admin. Report No. FE-1730-21. National Technical Information Service, Springfield, Va., 1976.
32. Mahajan, O. P., Tomita, A., Nelson, J. R., and Walker, P. L., Jr., *Am. Chem. Soc., Div. Fuel Chem., Prepr.*, 20(3), 19-22 (1975).
33. Lee, A. L., Feldkirchner, H. L., Schora, F. C., and Henry, J. J., *Ind. Eng. Chem., Proc. Des. Dev.*, 7, 244-249 (1968).

REFERENCES

34. Woebcke, H. N., Chambers, L. E., and Virk, P. S., *Am. Chem. Soc., Div. Fuel Chem., Prepr.*, **18**(2), 97-132 (1973).
35. Stephens, D. R., U.S. Atomic Energy Commission Report No. 16094, National Technical Information Service, Springfield, Va., 1972.
36. Hougen, O. A., and Watson, K. M., *Industrial Chemical Calculations*, 2nd ed., Wiley, New York, 1936, 487 pp.
37. van Krevelen, D. W., *Coal*, Elsevier, Amsterdam, 1961, 514 pp.
38. Janka, J. C., and Malhotra, R., U.S. Office Coal Research, R&D Report No. 22—Interim Report No. 7 (PB 235 370), National Technical Information Service, Springfield, Va., 1971.
39. Zenz, F. A., and Othmer, D. F., *Fluidization and Fluid-Particle Systems*, Reinhold, New York, 1960, 513 pp.
40. Field, M. A., Gill, D. S., Morgan, D. B., and Hawksley, P. G. W., *Combustion of Pulverized Coal*, British Coal Utilisation Research Association, Leatherhead, England, 1967, 413 pp.
41. Davidson, J. F., and Harrison, D., *Fluidised Particles*, Cambridge U. P., New York, 1963, 155 pp.
42. Murray, J. D., *Chem. Eng. Progr. Symp. Ser.*, **62**, 71-82 (1966).
43. Jackson, R., *Transact. Inst. Chem. Eng.*, **41**, 13-28 (1962).
44. Grace, J. R., *AIChE Symp. Ser.*, **67**(116), 159 (1971).
45. Pyle, D. L., *Advances in Chemistry Series*, No. 109, American Chemical Society, Washington, D.C., 1972, pp. 106-130.
46. Rowe, P. N., in *Proceedings of Fifth European/Second International Symposium on Chemical Reaction Engineering*, J. M. H. Fortuin, Ed., Elsevier, Amsterdam, 1972, pp. A9-1-14.
47. Chavarie, C., and Grace, J. R., *Ind. Eng. Chem., Fundam.*, **14**, 75 (1975).
48. Davidson, J. F., and Harrison, D., *Fluidization*, Academic Press, London, 1971.
49. Shen, C. Y., and Johnstone, H. F., *AIChE Journal*, **1**, 349-354 (1955).
50. Mathis, J. F., and Watson, C. C., *AIChE Journal*, **2**, 518-524 (1956).
51. Lewis, W. K., Gilliland, E. R., and Glass, W., *AIChE Journal*, **5**, 419 (1959).
52. May, W. G., *Chem. Eng. Progr.*, **55**(12), 49-56 (1959).
53. Lanneau, K. P., *Inst. Chem. Eng. Transact.*, **38**, 125-143 (1960).
54. Gomezplata, A., and Shuster, W. W., *AIChE Journal*, **6**, 454-459 (1960).
55. van Deemter, J. J., *Chem. Eng. Sci.*, **13**, 143-154 (1961).
56. Mamuro, T., and Muchi, I., *Internat. Chem. Eng.*, **5**, 732-736 (1965).
57. Kobayashi, H., Arai, F., Chiba, T., and Tanaka, Y., *Kagaku Kogaku*, **33**, 274-280 (1969).
58. Mireur, J. P., and Bischoff, K. B., *AIChE Journal*, **13**, 839-845 (1967).
59. van Deemter, J. J., in *Proc. International Symposium on Fluidization*, A. A. H. Drinkenburg, Ed., Netherlands U. Press, Amsterdam, 1967, pp. 334-347.
60. Orcutt, J. C., Davidson, J. F., and Pigford, R. L., *Chem. Eng. Progr. Symp. Ser.*, **58**(38), 1-15 (1962).
61. Rowe, P. N., *Chem. Eng. Progr.*, **60**(3), 75-80 (1964).
62. Rose, P. L., Ph.D. thesis, Cambridge, 1965.

172 FUNDAMENTALS OF COAL GASIFICATION

63. Partridge, B. A., and Rowe, P. N., *Transact. Inst. Chem. Eng.*, **44**, 335–348 (1966).
64. Toor, F. D., and Calderbank, P. H., in *Proceedings of International Symposium on Fluidization*, A. A. H. Drinkenburg, Ed., Netherlands U. P., Amsterdam, 1967, pp. 373–392.
65. Hovmand, S., and Davidson, J. F., *Transact. Inst. Chem. Eng.*, **46**, 190–203 (1968).
66. Kunii, D., and Levenspiel, O., *Fluidization Engineering*, Wiley, New York, 1969, 534 pp.
67. Kato, K., and Wen, C. Y., *Chem. Eng. Sci.*, **24**, 1351–1369 (1969).
68. Chiba, T., and Kobayashi, H., *Chem. Eng. Sci.*, **25**, 1375–1385 (1970).
69. Austin, L. G., Gardner, R. P., and Walker, P. L., Jr., *Fuel*, **42**, 319–323 (1963).
70. Leva, M., Weintraub, M., Grummer, M., and Pollchik, M., *Ind. Eng. Chem.*, **41**, 1206 (1949).
71. Perry, R. H., and Chilton, C. H., *Chemical Engineers' Handbook*, 5th ed., McGraw-Hill, New York, 1973.
72. Agarwal, O. P., and Storrow, J. A., *Chem. Ind. (Lond.)*, 278–286 (1951).
73. Leva, M., Weintraub, M., Grummer, M., Pollchik, M., and Storch, H. H., *U.S. Bur. Mines Bull.* (504), 1–149 (1951).
74. van Heerden, C., Nobel, A. P. P., and van Krevelen, D. W., *Chem. Eng. Sci.*, **1**, 37–49 (1951).
75. Jones, J. F., Eddinger, R. T., and Seglin, L., in *Proc. Internat. Symp. Fluidization*, A. A. H. Drinkenburg, Ed., Netherlands U. P., Amsterdam, 1967, pp. 676–687.
76. Curran, G. P., and Gorin, E., *U.S. Office Coal Res. R&D Report No., 16–Interim Rept. No. 3, Book 1* (PB 184718), National Technical Information Service, Springfield, Va., 1970.
77. Feldmann, H. F., Kiang, K., and Yavorsky, P., *Am. Chem. Soc., Div. Fuel Chem., Prepr.*, **15**(3), 62 (1971).
78. Leva, M., *Fluidization*, McGraw-Hill, New York, 1959, 327 pp.
79. Wen, C. Y., and Yu, Y. H., *AIChE Journal*, **12**, 610–612 (1966).
80. Frantz, J. F., *Chem. Eng.*, **69**(19), 161–178 (1962).
81. Ergun, S., *Chem. Eng. Progr.*, **48**, 89–94 (1952).
82. Knowlton, T. M., Paper No. 9b, presented at 67th Annual Meeting of American Institute of Chemical Engineers, Washington, D.C., December 1–5, 1974.
83. Tarman, P., Punwani, D., Bush, M., and Talwalkar, H., *U.S. Office Coal Res. R&D Rept. No. 95–Interim Report No. 1* (PB 237363), National Technical Information Service, Springfield, Va., 1975.
84. Bakker, P. J., and Heertjes, P. M., *Br. Chem. Eng.*, **3**, 240–245 (1958).
85. Bakker, P. J., and Heertjes, P. M., *Chem. Eng. Sci.*, **12**, 260–271 (1960).
86. Bituminous Coal Research Inc., *U.S. Energy Res. Develop. Admin. Rept. No. FE-1527-1*, National Technical Information Service, Springfield, Va., 1975.
87. Fink, C. E., Sudbury, J. D., and Curran, G. P., Paper No. 15d, presented at

77th National Meeting of American Institute of Chemical Engineers, June 2-5, 1974.
88. Talmor, E., and Benenati, R. F., *AIChE Journal*, **9**, 536-540 (1963).
89. Katz, S., and Zenz, F. A., *Hydrocarbon Process Petrol. Refiner*, **33**, 203 (1954).
90. Koble, R. A., Thesis, West Virginia University, Morgantown, W. Va., 1952.
91. Leva, M., and Grummer, M., *Chem. Eng. Progr.*, **48**, 307-313 (1952).
92. Woolard, I. N. M., and Potter, O. E., *AIChE Journal*, **14**, 388-391 (1968).
93. Bart, R., Ph.D. thesis, Massachusetts Institute of Technology, Cambridge, Mass., 1950.
94. Vakhrushev, I. A., and Erokhin, G. S., *Intern. Chem. Eng.*, **3**, 333-338 (1963).
95. Yoshida, K., and Kunii, D., *J. Chem. Eng. Jap.*, **1**, 11-16 (1968).
96. Anderson, R. B., Hall, W. K. Lecky, J. A., and Stein, K. C., *J. Phys. Chem.*, **60**, 1548-1558 (1956).
97. Gan, H., Nandi, S. P., and Walker, P. L., Jr., *U.S. Office Coal Res. R&D Report No. 61—Interim Rept. No. 4* (PB 233994), National Technical Information Service, Springfield, Va., 1972.
98. Gan. H., Nandi, S. P., and Walker, P. L., Jr., *Fuel*, **51**, 272-277 (1972).
99. Walker, P. L., Jr., and Hippo, E. J., *Am. Chem. Soc., Div. Fuel Chem., Prepr.*, **20**(3), 45-51 (1975).
100. Cartz, L., and Hirsch, P. G., *Trans. Royal Soc. (Lond.)*, **A252**, 557 (1960).
101. Walker, P. L., Jr., Austin, L. G., and Nandi, S. P., *Chemistry and Physics of Carbon*, Vol. 2, Marcel Dekker, New York, 1960, pp. 257-371.
102. Franklin, R. E., *Fuel*, **27**, 46-49 (1948).
103. Hirshfelder, J. G., Curtis, C. F., and Bird, R. B., *Molecular Theory of Gases and Liquids*, Wiley, New York, 1954, p. 1127.
104. Maggs, F. A. P., Schwabe, P. H., and Williams, J. H., *Nature (Lond.)*, **186**, 956-958 (1960).
105. Kini, K. A., and Stacy, W. O., *Carbon* **1**, 17-24 (1963).
106. Bond, R. L., *Porous Carbon Solids*, Academic Press, New York, 1967, p. 99.
107. Ergun, S., and Menster, M., *Fuel*, **39**, 509-510 (1960).
108. Agrawal, P. L., *Proceedings of Symposium on Nature Coal, Central Fuel Research Institute, Jealgora, India, 1959*, pp. 121-128.
109. Bond. R. L., and Spencer, D. H. T., *Soc. Chem. Ind., Lond. Chem. Eng. Group, Proc.*, 231-251 (1958).
110. Dormans, H. N. M., Huntjens, F. J., and van Krevelen, D. W., *Fuel*, **36**, 321-339 (1957).
111. Franklin, R. E., *Transact. Faraday Soc.*, **45**, 668-682 (1949).
112. Fjuii, S., and Tsuboi, H., *Fuel*, **46**, 361-366 (1967).
113. Fjuii, S., and Shinbata, T., *Nenryo Kyokai-Shi (Fuel Soc. Jap. J.)*, **45**, 868-875 (1966).
114. Osawa, Y., Sugimura, H., Kawaski. Y., and Fujii, S., *Nenryo Kyokai-Shi (Fuel Soc. Jap. J.)*, **49**, 31-39 (1970).

115. Sugimura, H., Osawa, Y., Hatami, M., Sato, H., and Honda, H., *Nenryo Kyokai-Shi* (*Fuel Soc. Jap. J.*), **45**, 618–630 (1966).
116. Wandless, A. M., and Macrae, J. C., *Fuel*, **13**, 4 (1934).
117. Toda, Y., *Report No. 5—Study on Pore Structure of Coals and Their Carbonized Products*, National Research Institute for Pollution and Resources, Kawaguchi, Japan, 1973, 112 pp. (in Japanese).
118. Washburn, E. W., *Proc. Nat. Acad. Sci.* (*USA*), **7**, 115 (1921).
119. Feldmann, H. F., Mima, J. A., and Yavorsky, P. M., *Adv. Chem. Ser.* (131), 108–125, (1974).
120. Lewellen, P. C., Ph.D. thesis, Massachusetts Institute of Technology, Cambridge, Mass., 1976.
121. Johnson, J. L., *Am. Chem. Soc., Div. Fuel Chem., Prepr.*, **20**(4), 85–101 (1975).
122. Torikai, N., and Walker, P. L., Jr., *U.S. Office of Coal Res. SROCR-3 Report*, Washington, D.C., 1968.
123. Hippo, E. J., Ph.D. thesis, The Pennsylvania State University, University Park, Pa., 1977.
124. Dutta, S., Wen, C. Y., and Belt, R. J., *Am. Chem. Soc., Div. Fuel Chem., Prepr.*, **20**(3), 103–114 (1975).
125. Stacy, W. O., and Walker, P. L., *U.S. Office Coal Res. R&D Report No. 61–Interim Rept. No. 3* (PB 233996), National Technical Information Service, Springfield, Va., 1973.
126. Brunauer, S., Emmett, P. H., and Teller, E., *J. Am. Chem. Soc.*, **60**, 309 (1938).
127. Lamond, T. G., and Marsh, H., *Carbon*, **1**, 281 (1964).
128. Everett, D. H., *Proc. Chem. Soc.* (*Lond.*), 38–53 (1957).
129. Marsh, H., and Siemieniewska, T., *Fuel*, **44**, 355 (1965).
130. Marsh, H., and Wynne-Jones, W. F. K., *Carbon*, **1**, 281 (1964).
131. Walker, P. L., and Patel, R. L., *Fuel*, **49**, 91 (1970).
132. Spencer, D. H. T., and Bond. R. L., *Advances in Chemistry Series,* No. 55, American Chemical Society, Washington, D.C., 1966, pp. 724–728.
133. Dubinin, M. M., in *Chemistry and Physics of Carbon*, Vol. 2, P. L. Walker, Jr., Ed., Marcel Dekker, New York, 1966, p. 51.
134. Ergun, S., and Mentser, M., in *Chemistry and Physics of Carbon*, P. L. Walker, Jr., Ed., Marcel Dekker, New York, 1965, pp. 204–263.
135. Dent, F. J., *Gas J.*, **244**, 502–505 (1944).
136. Birch, T. J., Hall, K. R., and Urie, R. W., *J. Inst. Fuel*, **33**, 422–435 (1960).
137. Blackwood, J. D., and McCarthy, D. J., *Aust. J. Chem.*, **19**, 797–813 (1966).
138. Feldkirchner, H. L., and Linden, H. R., *Ind. Eng. Chem., Process Des. Dev.*, **2**, 153–162 (1963).
139. Wen, C. Y., and Huebler, J., *Ind. Eng. Chem. Process Des. Dev.*, **4**, 142–147 (1965).
140. Moseley, F., and Patterson, D., *J. Inst. Fuel*, **38**, 378–391 (1965).
141. Moseley, F., and Patterson, D., *J. Inst. Fuel*, **40**, 523–530 (1967).

REFERENCES 175

142. Feldkirchner, H. L., and Huebler, J., *Ind. Eng. Chem. Process Des. Dev.*, 4, 134-142 (1965).
143. Hiteshue, R. W., Friedman, S., and Madden, R., *U.S. Bur. Mines, Rept. Invest.* (6376) (1964).
144. Anthony, D. B., thesis, Massachusetts Institute of Technology, Cambridge, Mass., 1974.
145. Anthony, D. B., Howard, J. B., Hottel, H. C., and Meissner, H. P., *Fuel*, 55, 121-128 (1976).
146. Anthony, D. B., and Howard, J. B., *AIChE Journal*, 22, 625-656 (1976).
147. Graff, R. A., Dobner, S., and Squires, A. M., *Am. Chem. Soc., Div. Fuel Chem., Prepr.*, 20(3), 23-32 (1975).
148. Graff, R. A., Dobner, S., and Squires, A. M., *Fuel*, 55, 109-115 (1976).
149. Growcock, F. B., and MacKenzie, D. R., *Fuel*, 55, 349-354 (1976).
150. Lewis, P. S., Friedman, S., and Hiteshue, R. W., *Advances in Chemistry Series*, No. 69, American Chemical Society, Washington, D.C., 1967, pp. 50-63.
151. Glenn, R. A., Donath, E. E., and Grace, R. J., *Advances in Chemistry Series*, No. 69, American Chemical Society, Washington, D.C., 1967, pp. 81-103.
152. Feldmann, H. F., Simons, W. H., Mima, J. A., and Hiteshue, R. W., *Am. Chem. Soc., Div. Fuel Chem., Prepr.*, 14(4), Part 2, 1-25 (1970).
153. Johnson, J. L., *Am. Chem. Soc., Div. Fuel Chem., Prepr.*, 20(3), 61-87 (1975).
154. Johnson, J. L., *Am. Chem. Soc., Div. Fuel Chem., Prepr.*, 22(1), 17-37 (1977).
155. Johnson, J. L., paper presented at Joint USA-USSR Symposium on Coal Gasification/Liquefaction, Moscow, October 1976.
156. Steinberg, M., and Fallon, P., paper presented at 68th Annual Meeting of American Institute of Chemical Engineers, Los Angeles, Calif., November 16-20, 1975.
157. Zahradnik, R. L., and Glenn, R. A., *Fuel*, 50, 77-90 (1971).
158. Bituminous Coal Research, Inc., *U.S. Office Coal Res. R&D Report No. 20—Final Report* (PB 235530), National Technical Information Service, Springfield, Va., 1970.
159. Pyrcioch, E. J., Feldkirchner, H. L., Tsaros, C. L., Johnson, J. L., Bair, W. G., Lee, B. S., Schora, F. C., Huebler, J., and Linden, H. R., *Inst. Gas Technol., Chicago, Res. Bull.*, (39), 225 pp. (1972).
160. Zahradnik, R. L., and Grace, R. J., *Advances in Chemistry Series*, No. 131, American Chemical Society, Washington, D.C., 1974, pp. 126-144.
161. Hill, G. R., Anderson, L. L., Wiser, W. H., Qadar, S. A., Wood, R. E., and Bodily, D. M., *U.S. Office Coal Res. R&D Report No. 18—Final Report* (PB 224748), National Technical Information Service, Springfield, Va., 1970.
162. Gray, J. A., Donatelli, D. J., and Yavorsky, P. M., *Am. Chem. Soc., Div. Fuel Chem., Prepr.*, 20(4), 103-145 (1975).
163. Zielke, C. W., and Gorin, E., *Ind. Eng. Chem.*, 47, 820-825 (1955).

164. Sastry, S., Mallikarjunan, M. M., and Vaidyeswaran, R., *Brennstoff-Chem.*, **50**, 363–367 (1969).
165. Blackwood, J. D., McCarthy, D. J., and Cullis, B. D., *Aust. J. Chem.*, **20**, 2525–2528 (1967).
166. Blackwood, J. D., Cullis, B. D., and McCarthy, D. J., *Aust. J. Chem.*, **20**, 1561–1570 (1967).
167. Blackwood, J. D., *Aust. J. Chem.*, **12**, 14–28 (1959).
168. Tomita, A., Mahajan, O. P., and Walker, P. L., Jr., *Am. Chem. Soc., Div. Fuel Chem., Prepr.*, **20**(3), 99–102 (1975).
169. Moseley, F., and Patterson, D., *J. Inst. Fuel*, **38**, 13–23 (1965).
170. Blackwood, J. D., *Aust. J. Chem.*, **15**, 397–408 (1962).
171. Barrer, R. M., *Proc. Roy. Soc. (Lond.)*, **A149**, 253–269 (1935).
172. Blackwood, J. D., *Aust. J. Appl. Sci.*, **13**, 199–206 (1962).
173. Jüntgen, H., Van Heek, K. H., and Klein, J., paper presented at Gordon Research Conference on Coal Science, New Hampton School, New Hampton, N. H., July 1973.
174. Zielke, C. W., and Gorin, E., *Ind. Eng. Chem.*, **49**, 396–403 (1957).
175. Walker, P. L., Jr., Austin, L. G., and Rusinko, F., Jr., *Adv. Catal.*, Vol. 11, Academic Press, New York, 1959, pp. 133–221.
176. Dent, F. J., Blackburn, W. H., and Millett, H. C., *Transact. Inst. Gas Eng.*, **88**, 150–217 (1938).
177. Mentser, M., and Ergun, S., *U.S. Bur. Mines Bull.* (644), 42 pp. (1973).
178. Blackwood, J. D., and Ingeme, A. J., *Aust. J. Chem.*, **13**, 194–209 (1960).
179. Fuchs, W., and Yavorsky, P. M., *Am. Chem. Soc., Div. Fuel Chem., Prepr.*, **20**(3), 115–133 (1975).
180. Yergey, A. L., and Lampe, F. W., *Fuel*, **53**, 280–281 (1974).
181. Marsh, H., and Taylor, D. W., *Fuel*, **54**, 218–219 (1975).
182. Taylor, R. W., and Bowen, D. W., Lawrence Livermore Laboratories Report, No. UCRL 52002, National Technical Information Service, Springfield, Va., 1976.
183. Dutta, S., Wen, C. Y., and Belt, R. J., *Am. Chem. Soc., Div. Fuel Chem., Prepr.*, **20**(3), 103–114 (1975).
184. Grabke, H. J., *Ber. Bunsenges. Phys. Chem.*, **70**, 664–674 (1966).
185. Hottel, H. C., Williams, G. C., and Wu, P. C., *Am. Chem. Soc., Div. Fuel Chem., Prepr.*, **10**(3), 58–71 (1966).
186. Biederman, D. L., Ph.D. thesis, The Pennsylvania State University, University Park, Pa., 1965, 105 pp.
187. Strange, J. F., Ph.D. thesis, The Pennsylvania State University, University Park, Pa., 1964, 120 pp.
188. Ergun, S., *U.S. Bur. Mines Bull.* (598), 38 pp. (1962).
189. Ergun, S., *J. Phys. Chem.*, **60**, 480–485 (1956).
190. Lewis, W. K., Gilliland, E. R., and McBride, G. T., Jr., *Ind. Eng. Chem.*, **41**, 1213–1226 (1949).
191. Gadsby, J., Long, F. J., Sleightholm, P., and Sykes, K. W., *Proc. Roy. Soc. (Lond.)*, **A193**, 357–376 (1948).
192. Duval, X., *J. Chim. Phys.*, **47**, 339–347 (1950).

REFERENCES

193. Duval, X., *Ann. chim.*, **10**, 903–967 (1955).
194. Spackman, W., Davis, A., Walker, P. L., Lovell, H. L., Essenhigh, R. H., Vastola, F. J., and Given, P. H., U.S. Energy Res. Develop. Admin. Report No. FE-2031-1, National Technical Information Service, Springfield, Va., 1975.
195. Hippo, E., and Walker, P. L., Jr., *Fuel*, **54**, 245–248 (1975).
196. Blackwood, J. D., *Coke and Gas*, **22**, 293–297 (1960).
197. Long, F. J., and Sykes, K. W., *Proc. Roy. Soc. (Lond.)*, **A193**, 377–399 (1948).
198. Blakely, J. P., and Overholser, L. G., *Carbon*, **3**, 269–275 (1965).
199. Gadsby, J., Hinshelwood, F. R. S., and Sykes, K. W., *Proc. Roy. Soc. (Lond.)*, **A187**, 129–151 (1946).
200. Blackwood, J. D., and McGrory, F., *Aust. J. Chem.*, **11**, 16–33 (1958).
201. Curran, G. P., Fink, C. E., and Gorin, E., *Am. Chem. Soc., Div. Fuel Chem., Prepr.*, **12**(3), 62–82 (1968).
202. Johnstone, H. F., Chen, C. Y., and Scott, D. S., *Ind. Eng. Chem.*, **44**, 1564–1569 (1952).
203. Strickland-Constable, R. F., *J. Chim. Phys.*, **47**, 356–360 (1950).
204. Key, A., *Commun. GRB 40*, Gas Research Board, London, 1948, 36 pp.
205. Cherednik, E. M., Apel'baum, L. O., and Temkin, M. I., *Doklady Akad. Nauk SSSR*, **174**(4), 891–894 (1967).
206. Temkin, M. I., Cherednik, E. M., and Apel'baum, L. O., *Kinet. Catal. (Engl. Transl.)*, **9**(1), Part 1, 76–83 (1968).
207. Goring, G. E., Curran, G. P., Tarbox, R. P., and Gorin, E., *Ind. Eng. Chem.*, **44**, 1057–1065 (1952).
208. Goring, G. E., Curran, G. P., Tarbo, R. P., and Gorin, E., *Ind. Eng. Chem.*, **45**, 2856–2891 (1953).
209. Van Heek, K. H., Jüntgen, H., and Peters, W., *J. Inst. Fuel*, **46**, 249–258 (1973).
210. Goring, G. E., Curran, G. P., Tarbox, R. P., and Gorin, E., *Ind. Eng. Chem.*, **44**, 1051–1057 (1952).
211. Blackwood, J. D., *Coke and Gas*, **22**, 190–218 (1960).
212. Hedden, K., and Mienkina, G., *Brennstoff-Chem.*, **46**, 366–371 (1965).
213. Curran, G., and Gorin, E., *U.S. Office Coal Res. R&D Rept. No. 16– Interim Rept. No. 3, Book 2* (PB 184719), National Technical Information Service, Springfield, Va., 1970.
214. Walker, P. L., Jr., Austin, L. G., and Rusinko, F., Jr., *Adv. Catal.*, Vol. 11, Academic Press, New York, 1959, pp. 133–221.
215. Kunio, Y., and Kunii, D., *J. Chem. Eng. Jap.*, **2**, 170–174 (1969).
216. Thiele, E. W., *Ind. Eng. Chem.*, **31**, 916 (1939).
217. Zeldowitsch, J. B., *Acta Physicochim. USSR*, **10**, 583 (1939).
218. Wheeler, A., *Adv. Catal.*, Vol. 3, Academic Press, New York, 1951, pp. 249–327.
219. Walker, P. L., Jr., Austin, L. G., and Nandi, S. P., *Chemistry and Physics of Carbon*, Vol. 2, Marcel Dekker, New York, 1960, pp. 257–371.

220. Walker, P. L., Jr., Shelef, M., and Anderson, R. A., *Chemistry and Physics of Carbon*, Vol. 4; Marcel Dekker, New York, 1968, pp. 287–383.
221. Dent, F. J., Blackburn, W. H., and Millett, H. C., *Transact. Inst. Gas Eng.*, **88**, 150–217 (1938).
222. Fox, D. A., and White, A. H., *Ind. Eng. Chem.*, **23**, 259 (1931).
223. Frank, F. H., and Meraikib, M., *Carbon*, **8**, 423–433 (1970).
224. Kislykh, V. I., and Shishakov, N. V., *Gaz Prom.*, **8**, 15–19 (1960).
225. Kroger, C., *Angew. Chem.*, **52**, 129–139 (1939).
226. Kroger, C., and Melhorn, G., *Brennstoff-Chem.*, **19**, 257–262 (1938).
227. Long, F. J., and Sykes, K. W., *Proc. Roy. Soc. (Lond.)*, **A215**, 100–110 (1952).
228. Neumann, B., Kroger, C., and Fingas, E., *Z. Anor. Allg. Chemie*, **197**, 321–418 (1931).
229. Taylor, H. S., and Neville, H. A., *J. Am. Chem. Soc.*, **43**, 2055–2071 (1921).
230. Tomita, A., and Tamai, Y., *J. Catal.*, **27**, 293–300 (1972).
231. Vastola, F. J., and Walker, P. L., *J. Chem. Phys.*, **58**, 20–24 (1961).
232. Vignon, L., *Ann. chim.*, **15**, 42–60 (1921).
233. Gardner, N., Samuels, E., and Wilks, K., *Am. Chem. Soc., Div. Fuel Chem., Prepr.*, **18**(2), 43–85 (1973).
234. Wilks, K., Samuels, E., and Gardner, N., thesis of K. Wilks, Case Western Reserve University, Cleveland, Ohio, 1974.
235. Gardner, N., paper presented at Symp. Catal. Conv. Coal, Pittsburgh, Pa., April 1975.
236. Forney, A. J., Haynes, W. P., Gasior, S. J., and Kenny, R. F., *Am. Chem. Soc., Div. Fuel Chem., Prepr.*, **19**(1), 111–122 (1974).
237. Haynes, W. P., Gasior, S. J., and Forney, A. J., *Am. Chem. Soc., Div. Fuel Chem., Prepr.*, **18**(2), 1–28 (1973).
238. Cox, J. L., and Sealock, L. J., Jr., *Am. Chem. Soc., Div. Fuel Chem., Prepr.*, **19**(1), 64–78 (1974).
239. Hoffman, E. J., Cox, J. L., Hoffman, R. W., Roberts, J. A., and Willson, W. G., *Am. Chem. Soc., Div. Fuel Chem., Prepr.*, **16**(2), 64–67 (1972).
240. Willson, W. G., Sealock, L. J., Jr., Hoodmaker, F. C., Hoffman, R. W., Cox, J. L., and Stinson, D. L., *Am. Chem. Soc., Div. Fuel Chem., Prepr.*, **18**(2), 29–42 (1973).
241. Johnson, J. L., *Catal. Rev.–Sci. Eng.*, **14**(1), 131–152 (1976).
242. Katzman, H., *Elec. Power Res. Inst.* Report No. EPRI-FF207-0-0, Palo Alto, Calif. (1974).
243. Mills, G. A., *Am. Chem. Soc., Div. Fuel Chem., Prepr.*, **16**(2), 107–123 (1972).

2 Kinetics of Coal Gasification in Hydrogen during Initial Reaction Stages

INTRODUCTION

For the last 15 years a strong incentive has existed in the United States to develop the technology for commercial gasification of coals to yield high-methane gas suitable as a substitute for natural gas. One of the general conclusions derived from the many experimental studies conducted is that the yield of light gaseous hydrocarbons obtained during initial coal heat up and for very short periods thereafter plays an important role in affecting the overall performances and thermal efficiencies of any gasification process. It is during this initial gasification stage that coals undergo devolatilization reactions leading to the formation of carbon oxides, water, oils, and tars and, most importantly, significant quantities of light hydrocarbons—particularly methane—in the presence of hydrogen at elevated pressures. However, since the exceptionally high reactivity most coals exhibit for methane formation during intial reaction stages is transient—existing only for a period of seconds at higher temperatures—rational design of commercial systems to optimize methane yields requires a kinetic characterization of the pertinent processes occurring that is as detailed as possible.

Because of its importance this reaction has been studied in a variety of experimental investigations using fixed beds [1, 2, 8-12, 17], fluidized beds [3, 4, 12, 17], and dilute solid-phase systems [5-7, 13-16, 19-20]. In spite of the extensive amount of information obtained from these studies, however, primary emphasis in the development of kinetic correlations has been placed on description of the total methane yields obtained after relative deactivation of coal solids has occurred rather than on the more detailed behavior occurring during the transient period of "rapid-rate" methane formation.

An experimental study was thus initiated at the Institute of Gas Technology

---

Adapted from a paper presented at the Joint U.S.-U.S.S.R. Symposium on Coal Gasification/Liquefaction, Moscow, U.S.S.R., October 12-13, 1976.

(IGT) to supplement existing information, with the objective of quantitatively characterizing intermediate reaction processes occurring prior to completion of the rapid-rate methane formation reaction. This chapter discusses some of the main results obtained thus far in our study of the gasification kinetics of low-rank coals in hydrogen, helium, and hydrogen–helium mixtures. The kinetic models developed to describe light gaseous hydrocarbon formation during initial reaction stages of the coals tested are also described and applied to selected results obtained in other studies directed toward gasification of high-volatility bituminous coals.

EXPERIMENTAL

A schematic diagram of the experimental apparatus used in the study is shown in Figure 1; the details of this system have been described previously, [13]. The main component of the system is a 1.6-mm-ID 60-m-long, helical coiled transport reactor. The reactor tube itself serves as the heating element with electrodes attached along the length of the coil to provide nine independent heating zones. This unique feature permits the establishment of temperature profiles along the length of the reactor tube that correspond to the desired constant heat-up rates of gas–solids mixtures passing through the tube. The reactor can also be operated isothermally. Coal particles (0.074–0.089-mm diameter) are mixed with feed gases at the top of the reactor coil, and the dilute gas–solids mixture (<0.1% solids by volume) passes through the reactor coil at temperatures very close to those imposed on the reactor wall. The char produced is collected in a sintered metal filter heated to 300°C for subsequent solids analyses. The gas then passes through a condenser system to collect water and condensable hydrocarbons. The dry gas is analyzed with a mass spectrograph (periodic samples) for major gaseous species and, in some tests, with a flame ionization detector for the concentration of total carbon in hydrocarbon species.

Steady-state conditions are achieved in less than 1 min after initiation of gas and coal flows, and sufficiently large gas:solid feed ratios are employed to result in less than 5% dilution of the dry exit gas with reaction products. Thus the partial pressures of feed-gas components remain essentially constant throughout the reactor tube. The main gaseous products measured include carbon monoxide (CO), carbon dioxide (CO_2), hydrogen (H_2), methane (CH_4), and ethane (C_2H_6). Other noncondensable gaseous hydrocarbons, particularly three- and four-carbon aliphatics and possibly some ethylene, were not usually accurately measured with the mass spectrograph due to the low concentrations present. The total carbon in this species group, referred to as C_3^+, was determined by difference—that is, by subtracting the carbon in methane and ethane from the total carbon in gaseous hydrocarbon species determined by flame ionization detec-

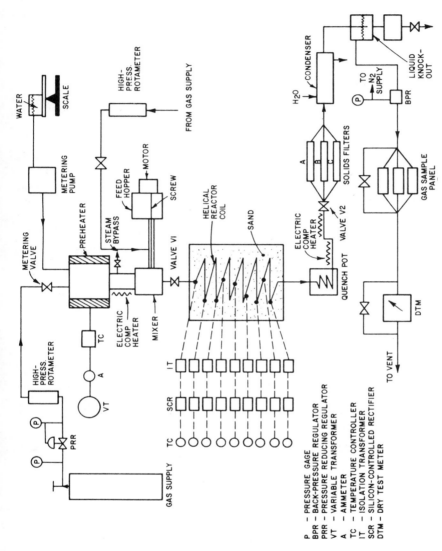

Figure 1 Diagram of the experimental system.

tion. Condensable liquid products were also difficult to quantitatively recover in the laboratory-scale equipment due to the relatively small quantities available. Thus product water was determined by an oxygen balance and condensable oils and tars, by a carbon balance.

TYPICAL RESULTS

The gasification kinetics of three low-rank coals—Montana lignite, Montana subbituminous coal, and North Dakota lignite—were investigated. The compositions of these three coals are given in Table 1, which also includes compositions of several other coals referred to in this chapter.

Figure 2 shows some typical product yields obtained for gasification of Montana lignite in hydrogen at 35 atm under conditions of a constant heat-up rate of 30°C/sec. With this lignite, devolatilization initiates at some temperature below 500°C. Coal oxygen is evolved primarily as carbon dioxide, carbon monoxide, and water, with a small amount observed in phenolic oils. Carbon dioxide evolution occurs below 500°C, apparently resulting from decomposition of carboxyl functional groups. Decomposition of other oxygenated functional groups leads to evolution of both carbon dioxide and water, with total available coal-oxygen evolution completed at about 700°C. This is shown more clearly in Figure 3. Although carbon dioxide yields begin decreasing above 650°C, this is due solely to the secondary reaction

$$CO_2 + H_2 \longrightarrow CO + H_2O \qquad [1]$$

occurring in the reactor coil, probably catalyzed by the reactor walls.

Oil and tar as well as C_3^+ gaseous hydrocarbon evolution occurs below 550°C. However, C_3^+ gaseous hydrocarbon yields decrease above 650°C due to hydrogenation to methane and ethane. Total yield of oil and tar remain relatively constant up to 850°C, although the composition of this fraction changes substantially with increasing temperature, becoming lighter in nature. Conversion of heavier oils to benzene initiates at about 650°C, and approximately one-half of the total oil/tar fraction consists of benzene at 850°C.

Methane-plus-ethane yields[1] increase continuously with increasing temperature, with rates of formation becoming substantial above 600°C. Above 700°C, with primary devolatilization processes virtually completed, gasification of the intermediate coal structure (semichar) is the primary reaction that occurs concurrently with the secondary devolatilization of coal hydrogen, leading to the

---

[1] Methane and ethane yields are grouped together since it has been previously shown [13] to be likely that, after initial devolatilization, both species are direct products of hydrogen gasification of the coal.

TABLE 1. COAL ANALYSES

	Montana Lignite	Montana Subbituminous (This Study)	North Dakota Lignite	Pennsylvania High-Volatile A Bituminous			Illinois High-Volatile C Bituminous (Ref. 1)	Illinois High-Volatile Bituminous (Ref. 2)	Warwickshire (Eng.) High-Volatile Bituminous (Ref. 5)
				(Ref. 1)	(Ref. 4)	(Ref. 3)			
Ultimate Analysis, wt% (MAF)									
Carbon	68.61	73.10	70.30	83.93	84.00	82.81	80.26	77.59	80.70
Hydrogen	4.35	4.53	4.47	5.54	5.70	5.69	5.44	5.80	5.86
Oxygen	25.50	19.73	23.60	7.66	7.10	7.70	11.07	10.58	13.44
Nitrogen	0.94	1.16	0.99	1.70	1.70	1.68	1.83	1.25	13.44
Sulfur	0.60	1.48	0.64	1.17	1.50	2.12	1.40	4.78	13.44
Total	100.00	100.00	100.00	100.00	100.00	100.00	100.00	100.00	100.00
Proximate Analysis, wt% (MAF)									
Volatile Matter	45.95	42.45	44.91	39.10	40.90	37.83	39.69	--	38.94
Fixed Carbon	54.05	57.55	55.09	60.90	59.10	62.17	60.31	--	61.06
Total	100.00	100.00	100.00	100.00	100.00	100.00	100.00	--	100.00

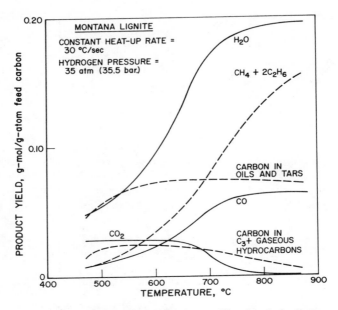

Figure 2 Typical product yields for gasification in hydrogen.

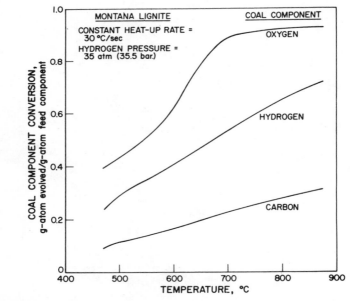

Figure 3 Typical coal-component conversion during gasification in hydrogen.

formation of relatively nonreactive coal char. Methane-plus-ethane yields increase significantly with increasing hydrogen pressure, (see Figure 4). This contrasts with the kinetics of evolution of primary devolatilization products, which depend primarily on time-temperature histories, and not on gas composition or pressure over the ranges of these variables employed.

The main qualitative features exhibited for gasification of Montana lignite in hydrogen at a heat-up rate of 30°C/sec were also observed for the other coals tested and for other time-temperature histories. As would be expected, yield-temperature curves such as those shown in Figure 2 are shifted somewhat toward higher temperatures at an increased heating rate of 80°C/sec and are shifted lower temperatures for isothermal operation with a gas-solids residence time of 7 sec. Significantly, however, the maximum yields of primary devolatilization products (carbon dioxide, carbon monoxide, water, oils and tars, and C_3^+ gaseous hydrocarbons) that can be achieved at sufficiently high temperatures appear to depend solely on coal type rather than on pressure, gas composition, or time-temperature history. This behavior is probably typical only of noncaking low-

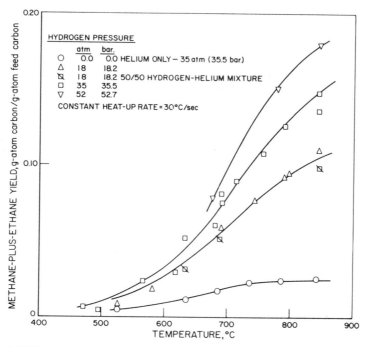

Figure 4 Effect of pressure on total methane-plus-ethane yields during gasification of Montana lignite.

rank coals. Different behavior would be expected with caking bituminous coals, particularly with respect to maximum oil and tar yields, which tend to increase with increased heat-up rate, decreased coal particle size, and decreased total pressure.

DATA ANALYSIS

A primary objective of this investigation is to quantitatively describe the kinetics of the initial-stage methane-plus-ethane formation for different coals as a function of hydrogen partial pressure and coal time-temperature history. The total methane-plus-ethane yields obtained in the gasification of coals with hydrogen result primarily from three overall reaction processes: direct coal hydrogenation, thermally activated coal-decomposition reactions, and secondary hydrogenation of C_3^+ gaseous hydrocarbons. Although some methane and ethane could also be formed by hydrogenation of light oils such as benzene, these reactions should not occur significantly at the conditions employed in this study [18].

Methane and Ethane From Coal-decomposition and Secondary Hydrogenation Reactions

To isolate the kinetics of direct coal hydrogenation, it is necessary to determine the contributions of coal-decomposition reactions and secondary hydrogenation reactions to total methane-plus-ethane yields. Since a fundamental approach to such evaluations would be extraordinarily complex, if practically possible at all, simplifying approximations and assumptions were made to obtain first-order estimates. Figure 5 shows methane-plus-ethane yields obtained for gasification of Montana lignite in helium at 35 atm, or 35.5 bar, for a constant heat-up rate of 30°C/sec and for isothermal operation with a coal-gas residence time of 7 sec. Based on these data, the methane-plus-ethane obtained from coal decomposition was assumed to occur in two stages: below 600°C a fixed fraction is instantaneously evolved; and above this temperature, yields increase linearly with increasing temperature up to 780°C when conversion is complete. The assumption that the kinetics of this reaction are independent of coal residence time implies a wide distribution of activation energies for the decomposition steps leading to methane-plus-ethane formation. It was assumed that the kinetics of this decomposition reaction in a helium atmosphere also apply in a hydrogen atomsphere.

Figure 6 shows yields of C_3^+ gaseous hydrocarbons obtained in hydrogen and helium for gasification of Montana lignite at a coal heat-up rate of 30°C/sec. In helium the C_3^+ hydrocarbons initially formed do not undergo pyrolysis at higher temperatures up to 810°C, whereas in hydrogen the yield of this fraction begins to decrease above about 600°C due to hydrogenation to methane and

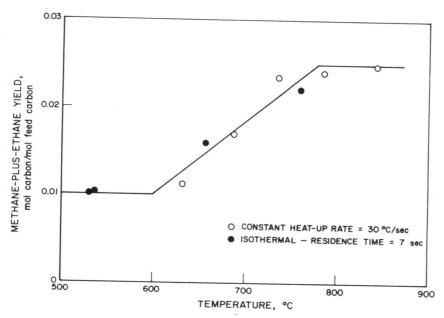

Figure 5 Methane-plus-ethane yields from pyrolysis of Montana lignite in helium at 35 atm (35.5 bar).

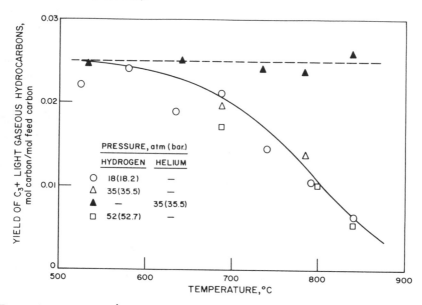

Figure 6 Yields of C_3^+ light gaseous hydrocarbon from gasification of Montana lignite in hydrogen or helium (constant gas–coal heat-up rate = 30°C/sec).

ethane. The overall hydrogenation reactions occurring were assumed to be equivalent to a single first-order reaction:

$$C_3^+ \text{ Gaseous hydrocarbons} \xrightarrow{k_4} \text{methane + ethane} \qquad [2]$$

The complete correlation to describe methane-plus-ethane formation from coal thermal decomposition and hydrogenation of C_3^+ hydrocarbons has the following form:

$$N = N_0 \, [\alpha' X' + \alpha'' X'' + \alpha''' X'''] \qquad [1]$$

where

X' = conversion fraction of thermal decomposition reactions below 600°C

X'' = conversion fraction of thermal decomposition reactions above 600°C

X''' = conversion fraction of C_3^+ hydrogenation reactions

N_0 = total carbon in methane-plus-ethane that can be formed from thermal decomposition and C_3^+ hydrogenation reactions (in g-atom/g-atom feed carbon in coal)

N = carbon in methane-plus-ethane formed from thermal decomposition and C_3^+ hydrogenation reactions at any time (in g-atom/g-atom feed carbon in coal)

$\alpha', \alpha'', \alpha'''$ = coefficients defining the distribution of carbon that can be converted to methane-plus-ethane from the three reactant groups considered

Based on data obtained with the three low-rank coals used in this study, and on some limited data available in the literature for gasification of a bituminous coal in hydrogen, the following numerical evaluations were made:

$$\alpha' = 0.2$$
$$\alpha'' = 0.3$$
$$\alpha''' = 0.5$$

	N_0, g-atom Carbon/g-atom Feed Carbon in Coal
Lignite	0.05
Subbituminous coal	0.08
Bituminous coal	0.10

$$X' = 1$$

$$X'' = 0 \text{ at } T < 873°K$$
$$X'' = 0.00555T - 4.845 \text{ for } 873° \leqslant T \leqslant 1053°K \qquad [2]$$
$$X'' = 1 \text{ for } T > 1053°K$$

and

$$\frac{dX'''}{d\theta} = k_4 (1 - X''')$$

$$k_4 = 1.7(10^4) \exp\left(\frac{-11,830}{T}\right) (\sec^{-1}) \qquad [3]$$

where

T = temperature (°K) in Equations 2 and 3
θ = time, (in sec)

Although the model described here is much oversimplified, it does provide an approximate basis consistent with the data obtained to numerically adjust methane-plus-ethane yields to correspond only to direct coal hydrogenation.

Kinetics of Direct Coal Hydrogenation

In a previous paper [13] a model was proposed to describe the kinetics of the initial stages of direct coal hydrogenation, based only on data obtained with Montana lignite and using a different method to adjust methane-plus-ethane yields than is described in the preceding section. These data were reevaluated in the context of all the data available at the present time, using the same model assumed previously but with improved and more genearlized evaluations of kinetic parameters.

A main feature leading to the development of the model is the strong correlation that exists between adjusted methane-plus-ethane yields and coal–hydrogen evolution during the secondary devolatilization stage. This behavior is shown in Figure 7, where methane-plus-ethane yields at constant hydrogen partial pressure are directly proportional to the secondary coal hydrogen evolved, independent of time–temperature history. Conveniently, the trends exhibited at a specific pressure level cannot be differentiated for the different coals considered. The slopes of the lines drawn in Figure 7 increase with increasing hydrogen pressure, consistent with qualitative observations made in other investigations of the rapid-rate methane phenomenon.

Variations in coal hydrogen evolution with increasing temperature are described in Figure 8. For a given time–temperature history the amount of coal

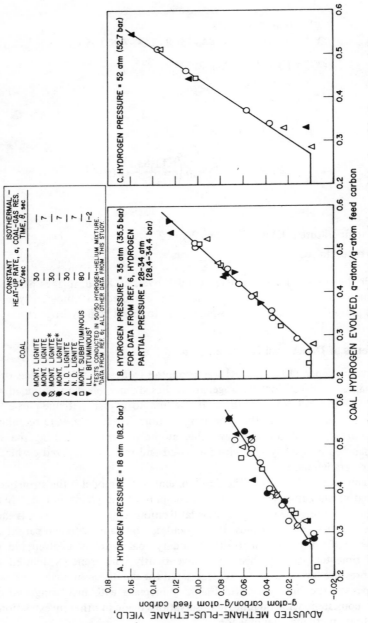

Figure 7 Stoichiometric relationship between adjusted methane-plus-ethane yield and coal-hydrogen evolution.

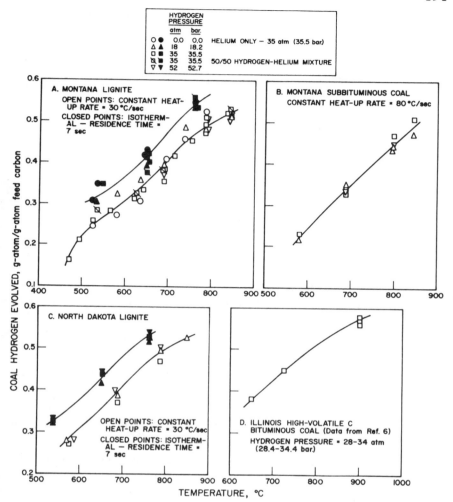

Figure 8 Coal-hydrogen evolution during gasification of different coals.

hydrogen evolved is essentially independent of gaseous hydrogen pressure, and the results obtained with Montana lignite (Figure 8a) show similar behavior in both helium and hydrogen atmospheres.

The results given in Figures 7 and 8 suggest that the formation of active sites, which promote methane-plus-ethane formation through interaction of gaseous hydrogen with the coal, is directly related to the processes in which coal hydrogen is evolved; this latter process involves only thermally activated phenomena occurring independently of gaseous atmosphere and dependent only on time-temperature history. A formal representation of this model is given by the following reactions:

$$CH_x^0 \xrightarrow[\text{(thermally activated coal-hydrogen evolution)}]{k_0} CH_y^* \quad [3]$$

Semichar   Active intermediate

$$CH_y^* \begin{cases} \xrightarrow{k_1} (1-m)CH_y^0 \text{ Inactive char} \\ \xrightarrow[\text{(gaseous hydrogen)}]{k_2} m\left[zCH_4 + \frac{(1-z)}{2}C_2H_6\right] \end{cases} \quad [4]$$

where

x = hydrogen:carbon atomic ratio in CH_x^0

y = hydrogen:carbon atomic ratio in CH_y^* and CH_y^0

z = fraction of carbon gasified as methane in Reaction 4

k_0, k_1, k_2 = first-order rate constants (sec^{-1})

$m = k_2/(k_1 + k_2)$

For purposes of quantitative correlation, the following definitions are made:

Y = total carbon in adjusted methane-plus-ethane yields

n_H = total coal hydrogen gasified at any time (in g-atom/g-atom feed carbon)

n_H^0 = total coal hydrogen gasified during primary devolatilization (in g-atom/g-atom feed carbon)

λ = fraction of total feed carbon not evolved as primary devolatilization products (closely related to fixed carbon) (in g-atom/g-atom feed carbon)

With this model the rate of Reaction 3 is assumed to be limiting, with the rate of Reaction 4 being very fast. In addition the ratio k_2/k_1 is assumed to be temperature independent. From these definitions the conversion fraction, f, of the semichar CH_x^0 to active intermediate CH_y^* can be expressed by the relationship

$$f = \frac{Y}{m\lambda} = \frac{n_H - n_H^0}{(x - y + my)\lambda} \quad [4]$$

From Equation 4 of the methane-plus-ethane yield, Y, is related to coal hydrogen evolved by

$$Y = \frac{m(n_H - n_H^0)}{(x - y + my)} \quad [5]$$

or

$$Y = S(n_H - n_H^0) \quad [6]$$

where

$$S = \frac{m}{(x - y + my)} \quad [7]$$

Equation 6 is consistent with the trends given in Figure 7. The slopes, S, of the lines drawn in Figure 7 are tabulated in Table 2, and the value of n_H^0 estimated from Figure 7 is about 0.27, corresponding to the value of coal hydrogen evolved due to primary devolatilization reactions.

The following independent relationship also results from the model assumed:

$$Y\left(\frac{1}{S - Z}\right) = \lambda(x - Z) \quad [8]$$

where Z is the hydrogen:carbon atomic ratio in the coal at a given time during gasification.

Using values of S given in Table 2, at appropriate hydrogen pressures, values of the term $Y(1/S - Z)$ were plotted against Z for the three coals tested and for the bituminous coal study to yield values of λ and x. A value of $\lambda = 0.8$ was applicable for all the coals considered. A value of $x = 0.65$, obtained for the bituminous coal, was slightly higher than a value of $x = 0.62$, obtained for the two

TABLE 2. VARIATION IN S WITH HYDROGEN PRESSURE

Hydrogen Pressure [atm (bar)]	S (g-atom Carbon g-atom Hydrogen)
18(18.2)	0.24
35(35.5)	0.41
52(52.7)	0.56

lignites and the subbituminous coal. With these evaluations, values of m were computed as a function of pressure by rearranging Equation 7:

$$m = \frac{S(x-y)}{1-Sy} \quad [9]$$

with a value of $y = 0.1$ assumed as a nominal average value based on gasification data obtained with different coals at elevated temperatures for extended times. Values of m and corresponding values of k_2/k_1 are given in Table 3, and values of k_2/k_1 are plotted against hydrogen pressure in Figure 9. The linear relationship shown in Figure 9 is represented by the equation

$$\frac{k_2}{k_1} = 0.0083 P_H \quad [10]$$

where P_H is hydrogen partial pressure, (in atmospheres)

From the evaluations made, the adjusted methane-plus-ethane yield is related to the conversion fraction, f, of the semichar by the relationship

$$Y = m\lambda f = \frac{0.00664 P_H}{1 + 0.0083 P_H} \cdot f \quad [11]$$

In analyzing the kinetics of Reactions 3 to quantitatively define the functional dependence of f on reaction conditions, the same model described in a previous paper was adopted. The main assumptions of this model are as follows:

- The compound CH_x^0 reacts according to Reaction 3 by a first-order process, but where there is a distribution of activation energies for the first-order rate constant, k_0.

TABLE 3. VARIATION OF m AND k_2/k_1 WITH HYDROGEN PRESSURE

Hydrogen Pressure [atm (bar)]	m	k_2/k_1
18(18.2)	0.127	0.145
35(35.5)	0.225	0.290
52(52.7)	0.307	0.443

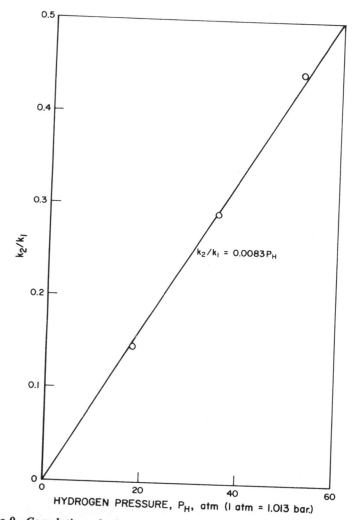

Figure 9 Correlation of relative rate ratio, $k_2 k_1$, with hydrogen pressure.

- The distribution function of activation energies is a constant; that is, $f(E)\,dE$ = fraction of total carbon in which the activation energy, E, in the rate constant

$$k_0 = k_0^0 \exp\left(\frac{-E}{RT}\right)$$

where

$f(E) = 0$ for $E < E_0$
$f(E) = C$ (constant) for $E_0 \leq E \leq E_1$
$f(E) = 0$ for $E > E_1$
k_0^0 = preexponential factor (sec^{-1})

is between E and $(E + dE)$.

Note that because

$$\int_{E_0}^{E_1} C\,dE = 1$$

then

$$C = \frac{1}{E_1 - E_0}$$

From these assumptions, the average conversion fraction of CH_x^0 can be expressed by the following relationship for any time–temperature history:

$$1 - f = \frac{1}{E_1 - E_0} \int_{E_0}^{E_1} \left\{\exp\left[-k_0^0 \int_0^\theta \exp\left(\frac{-E}{RT}\right) d\theta\right]\right\} dE \quad [12]$$

Statistical procedures were used to evaluate the parameters k_0^0, E_0, and E_1 given in the notation that follows, based on best consistency of experimental values of f computed from Equation 4 and calculated values of f computed from Equation 12 for the appropriate time–temperature history:

$k_0^0 = 1.97(10^{-10})$ (sec^{-1})
$E_0 = 40.8$ kcal/g-mol
$E_1 = 62.9$ kcal/g-mol

DATA ANALYSIS 197

A comparison between experimental and calculated values of f is given in Figure 10. A comparison of experimental total methane-plus-ethane yields and values calculated from the models and numerical evaluations presented in this chapter is given in Figure 11.

Although it may be tempting to consider theoretical implications concerning the detailed mechanisms of coal-hydrogen evolution during secondary

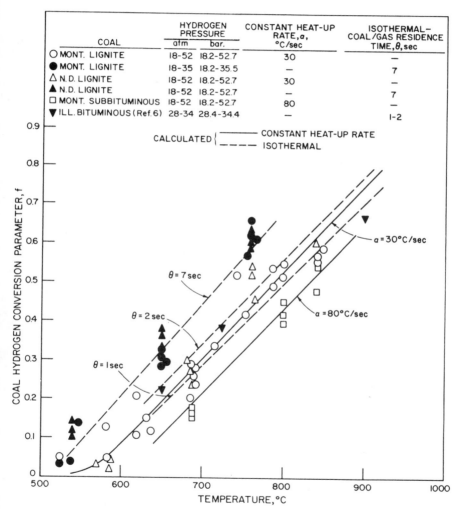

Figure 10 Comparison of calculated and experimental coal-hydrogen conversion.

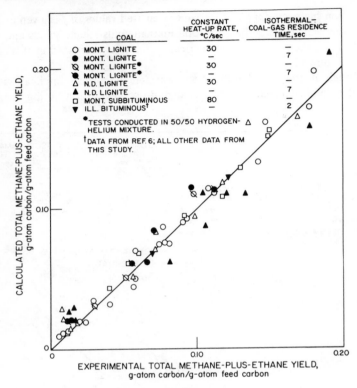

Figure 11 Comparison of calculated and experimental total methane-plus-ethane yields.

devolatilization, based on the values of k_0, E_0, and E_1 given in the preceding notation, such considerations should be made cautiously. Because of the sensitivity of these evaluations to the statistical numerical procedures employed and the strong covariance between these parameters, significantly different evaluations can be made to reasonably describe data obtained over a wide range of conditions. This point is illustrated in Figure 12, where variations in values of conversion, f, that were calculated using values of the kinetic parameters determined in this study are compared to values of f calculated using a significantly different set that nevertheless describes the same data quite well. This is particularly true for coal residence times ranging 1-10 sec, which represent the range of conditions most commonly employed in various investigations of initial coal gasification kinetics. The plot in Figure 12 was constructed to show the magnitude of differences in evaluation of kinetic parameters that could occur in describing data obtained over a wide range of residence times and

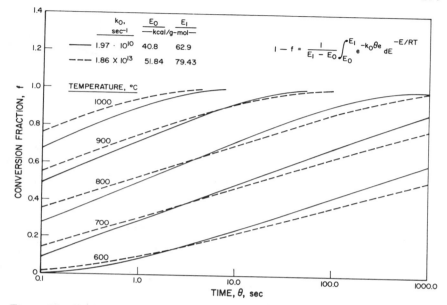

Figure 12 Comparison of model characteristics with different sets of kinetic parameters.

temperatures for isothermal conditions. Much greater ranges of estimates could result, however, in describing conversions obtained at a single residence time over a wide range of temperatures or, alternatively, at a single temperature over a wide range of residence times. Thus although the model proposed does suggest a mechanistic relationship between coal-hydrogen evolution and active site formation for coal hydrogenation, the numerical evaluations made to facilitate practical application of the information obtained for reactor design do not provide a substantial basis for a more detailed mechanistic understanding.

Application of Model to Bituminous Coal Gasification

The experimental study described has been limited to use of low-rank coals to minimize plugging of the small-diameter reactor tube employed. It was of interest, however, to make an approximate analysis of some data obtained in other investigations concerned with gasification of high-volatility-bituminous coals in hydrogen-containing gases to obtain an indication of the possible applicability of the model developed to describe behavior with bituminous coals. Figure 13 shows yield of methane-plus-ethane obtained in a variety of studies conducted over a wide range of conditions. Indicated values of coal and gas

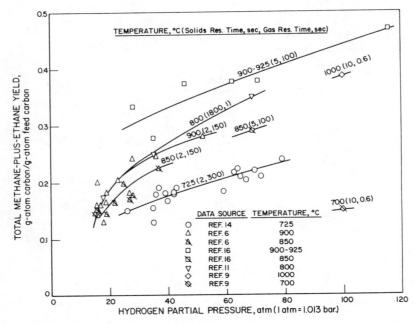

Figure 13 Total methane-plus-ethane yields obtained in different studies for gasification of bituminous coals with hydrogen.

residence times and hydrogen partial pressures were either reported directly or were deduced from other reported information. The hydrogen partial pressure used for correlation corresponds to the product gas. For systems where significant conversion of the feed gas occurs, which is the case in most of the systems considered, this representation assumes substantial backmixing in the gasification reactor. In all of the studies, either hydrogen or hydrogen–methane mixtures were used as feed gases. The analyses of the feed coals used in the various studies are included in Table 1.

In interpreting the data given in Figure 13, the model developed in this study was assumed. According to this model, experimental values of the function m should vary with hydrogen pressure according to the expression

$$m = \frac{0.0083 P_H}{1 + 0.0083 P_H} \qquad [13]$$

Values of m were computed from experimental values of adjusted methane-plus-ethane yield, Y, using the expression $m = (Y/\lambda f)$, where $\lambda = 0.8$. Values of Y

were computed by subtracting the values of N computed from Equation 1 from the experimental values of total methane-plus-ethane yield. Evaluations of N were made for the appropriate reported gas-solids residence times and reaction temperatures, using a value of $N_0 = 0.10$ presented previously for bituminous coal. Values of f were computed using Equation 9 (with values of k_0^0, E_0, and E_1 developed in this study), also at the reported reaction conditions.

In Figure 14 the m values computed from experimental results obtained with bituminous coals are compared to Equation 13. For hydrogen partial pressures up to about 120 atm, reasonable consistency exists, suggesting that the model developed is applicable for gasification of bituminous coals as well as for gasification of subbituminous coals and lignites. It should be noted, however, that in one study conducted with bituminous coal, [16] methane-plus-ethane yields higher than those predicted by the model were obtained for hydrogen partial pressures ranging 200–350 atm (202.6–354.6 bar). Although it is possible that some of this excess conversion might be due to methanation of carbon

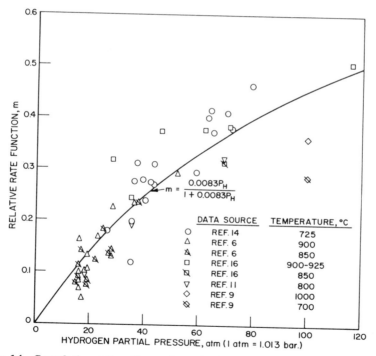

Figure 14 Correlation of methane-plus-ethane yields obtained in various investigations for high-volatility bituminous coal gasification in hydrogen.

oxides, it is also possible that the correlation defining hydrogen-pressure effect on the term k_2/k_1 should be adjusted to better describe results at very high pressures. At the present time, however, gasification at these pressure levels is of little practical interest, and available data are inadequate to justify a meaningful correlation.

SUMMARY

The results obtained in this study indicate that the initial gasification of low-rank coals in hydrogen-containing gases at elevated pressures occurs in two stages. The first stage involves thermally activated devolatilization reactions resulting in the evolution of carbon oxides, water, oil, tar, and some light gaseous hydrocarbons, and leading to the formation of an intermediate semichar. The second stage occurs consecutive to primary devolatilization and involves thermally activated decomposition reactions associated with the secondary devolatilization of remaining coal hydrogen, leading to the formation of a relatively nonreactive char. Significantly, yields of methane-plus-ethane, other than that derived from thermal coal-decomposition reactions or from C_3^+ light hydrocarbon hydrogenation reactions, are stoichiometrically related to coal hydrogen evolved during the secondary devolatilization stage. This has suggested a model in which the transitions that occur in conversion of semichar to char, with associated coal hydrogen evolution, involve the formation of a reactive intermediate carbon structure that can either rapidly react with gaseous hydrogen to form methane and ethane or can convert to nonreactive char. The ratio of methane-plus-ethane formation to char formation is independent of temperature but is directly proportional to hydrogen partial pressure.

A numerical representation of this model permits prediction of the kinetics of methane-plus-ethane formation for practical application to reactor design for systems using low-rank coals. A strong possibility that this model might also be suitable for application to gasification of high-volatility bituminous coal was also suggested, based on analyses of some available data obtained in other investigations.

REFERENCES

1. Anthony, D. G., "Rapid Devolatilization and Hydrogasification of Pulverized Coal," Sc.D. thesis, Massachusetts Institute of Technology, Cambridge, Mass., 1974.
2. Anthony, D. G., *Fuel*, 55(2), 121–128 (1976).
3. Birch, T. J., Hall, K. R., and Urie, R. W., *J. Inst. Fuel*, 33, 422–435 (1960).
4. Blackwood, J. D., and McCarthy, D. J., *Aust. J. Chem.*, 19, 797–813 (1966).

REFERENCES

5. Feldmann, H. F., "Reaction Model for Bituminous Coal Hydrogenation in a Dilute Phase," paper presented at the 160th National Meeting of the American Chemical Society, Division of Petroleum Chemistry, Chicago, September 13-18, 1970.
6. Feldmann, H. F., Mima, J. A., and Yavorsky, P. M., "Pressurized Hydrogasification of Raw Coal in a Dilute-Phase Reactor," *Coal Gasification*, in *Advances in Chemistry Series*, No. 131, American Chemical Society, Washington, D.C., 1974.
7. Glenn, R. A., Donath, E. E., and Grace, R. J., "Gasification of Coal Under Conditions Simulating Stage 2 of the BCR Two-Stage Super-Pressure Gasifier," *Fuel Gasification*, in *Advances in Chemistry Series*, No. 69, American Chemical Society, Washington, D.C., 1967.
8. Graff, R. A., Dobner, S., and Squires, A. M., "Products of Flash Hydrogenation," in *Proceedings of the 170th National Meeting of the American Chemical Society, Division of Fuel Chemistry*, Vol. 20, No. 3, 1975, pp. 23-32.
9. Graff, R. A., Dobner, S., and Squires, A. M., *Fuel*, 55(2), 109-115 (1976).
10. Hiteshue, R. W., Friedman, S., and Madden, R., "Hydrogasification of Bituminous Coals, Lignite, Anthracite, and Char," *U.S. Bur. Mines, Rept. Invest.* (6125) (1962).
11. Hiteshue, R. W., Friedman, S., and Madden, R., "Hydrogasification of a High-Volatile A Bituminous Coal," *U.S. Bur. Mines, Rept. Invest.* (6376) (1964).
12. Johnson, J. L., "Kinetics of Bituminous Coal Char Gasification With Gases Containing Steam and Hydrogen," *Coal Gasification*, in *Advances in Chemistry Series*, No. 131 American Chemical Society, Washington, D.C., 1974.
13. Johnson, J. L., "Gasification of Montana Lignite in Hydrogen in a Continuous Fluidized Bed Reactor," in *Proceedings of the 170th National Meeting of the American Chemical Society, Division of Fuel Chemistry*, Vol. 20, No. 3, 1975, pp. 61-87.
14. Lewis, P. F., Friedman, S., and Hiteshue, R. W., "High Btu Gas by Direct Conversion of Coal," *Fuel Gasification*, in *Advances in Chemistry Series*, No. 69, American Chemical Society, Washington, D.C., 1967.
15. Mosely, F., and Patterson, D., *J. Inst. Fuel*, 38, 378-391 (1965).
16. Mosely, F., and Patterson, D., *J. Inst. Fuel*, 40, 523-530 (1967).
17. Pyrcioch, E. J., Feldkirchner, H. L., Tsaros, C. L., Johnson, J. L., Bair, W. G., Lee, B. S., Schora, F. C., Huebler, J., and Linden, H. R., "Production of Pipeline Gas by Hydrogasification of Coal," *IGT Research Bulletin*, No. 39. Chicago, (November 1972).
18. Woebcke, H. N., Chambers, L. E., and Virk, P. S., "Thermal Synthesis and Hydrogasification of Aromatic Compounds," *Coal Gasification*, in *Advances in Chemistry Series*, No. 131, American Chemical Society, Washington, D.C., 1974.
19. Zahradnik, R. L., and Glenn, R. A., *Fuel*, 50, 77-90 (1971).
20. Zahradnik, R. L., and Grace, R. J., "Chemistry and Physics of Entrained Coal Gasification," *Coal Gasification*, in *Advances in Chemistry Series*, No. 131, American Chemical Society, Washington, D.C., 1974.

3 Gasification of Montana Lignite in Hydrogen and Helium during Initial Reaction Stages

INTRODUCTION

Light hydrocarbon yields obtained during the initial stages of coal gasification are of particular importance in affecting overall performances and thermal efficiencies of processes directed toward conversion of coal to pipeline gas. It is during this gasification stage that coals undergo devolatilization reactions leading to the formation of carbon oxides, water, oils, and tars, and, most importantly, significant quantities of light hydrocarbons, particularly methane, in the presence of hydrogen at elevated pressures. However, since the exceptionally high reactivity most coals exhibit for methane formation during initial reaction stages is transient, existing only for a period of seconds at higher temperatures, rational design of commercial systems to optimize methane yields requires as detailed a kinetic characterization of pertinent processes occurring as is possible. Because of its importance, this reaction has been studied in a variety of experimental investigations, using fixed beds [1, 5-7, 10], fluidized beds [2, 7, 10], and dilute solid-phase systems [3, 4, 8, 9, 11, 12]. In spite of the extensive amount of information obtained from these studies, however, primary emphasis in the development of kinetic correlations has been placed on description of total methane yields obtained after relative deactivation of coal solids has occurred, rather than on the more detailed behavior occurring during the transient period of "rapid-rate" methane formation. Although this existing information is of significant value at one level of process design, it is primarily limited to application to large-scale systems in which reaction conditions closely parallel the laboratory conditions employed in obtaining the information.

This current investigation has, therefore, been stimulated by the need for

---

Adapted from a paper presented at the 170th National Meeting of the ASC/DFC, Chicago, August 24-29. 1975, Fuel Chemistry Division, ACS, Preprints 20 (3): 61-87 (1975), with permission.

additional information that quantitatively characterizes intermediate reaction processes occurring prior to completion of the "rapid-rate" methane formation reaction. In this study, a continuous dilute-phase transport reactor has been employed having the particularly unique feature of variable temperature control along the length of the reactor, which permits the establishment of various desired gas-solid, time-temperature histories.

This chapter discusses some initial results obtained with this experimental system for gasification of Montana lignite in hydrogen, helium, and hydrogen-helium mixtures, under the more conventional conditions of isothermal operation, and under conditions of constant gas-solid heat-up rate ($\sim 50°F/sec$). Results are reported for tests conducted at temperatures of 900-1550°F and pressures of 18-52 atm.

EXPERIMENTAL

Apparatus and Procedures

The composition of the Montana lignite used in this study is given in Table 1, and a schematic diagram of the experimental apparatus is shown in Figure 1. The main component of the experimental system is a helical-coiled transport reactor formed from a $\frac{1}{16}$-in.-ID (inner diameter) tube. General information describing the reactor coil is given in Table 2. The diameter of the coil is about 1 ft, with a total tube length of 200 ft and a vertical reactor height of about $2\frac{1}{2}$ ft. With this design, gas flow rates of 5-50 scf/hr and solids flow rates of

TABLE 1 FEED COMPOSITION
(Montana Lignite, Dry Basis)

	Mass (%)
Ultimate analysis	
Carbon	65.13
Hydrogen	4.13
Oxygen	24.20
Nitrogen	0.89
Sulfur	0.57
Ash	5.08
	100.00
Proximate analysis	
Fixed Carbon	51.30
Volatile Matter	43.62
Ash	5.08
	100.00

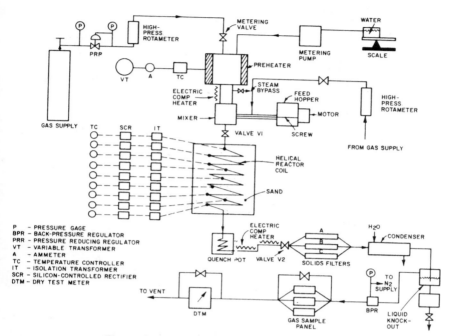

Figure 1 Diagram of the experimental system.

50–500 gph are possible. The relative gas–solids flow rates used in individual tests were such that solids:gas volume ratios were less than 0.02. The solids particles used in this system were relatively uniform in size, ranging 0.0029–0.0035 in. in diameter. Such small particles flowed essentially at gas velocities, and calculated temperature differences between the gas and solids and between the reactor tube wall and the flowing gas–solids stream were negligible. The reactor tube itself serves as the heating element, and electrodes are attached directly at various points along the length of the helical coil. Nine independent heating zones are thereby incorporated into the system to provide flexibility in establishing desired temperature profiles.

In a typical experimental test, the following operational procedures were generally used. Initially, the system is brought to a desired pressure, and a preliminary temperature profile is established in the reactor coil by adjusting the controls for the nine heating zones. When feed-gas flow is established at a desired rate, the flow from the solids feed hopper is initiated. Solids are screwed into a mixing zone, combining there with the feed gas, and the resulting mixture then flows through the reactor coil. The temperature in the mixing zone is maintained equal to the temperatue at the entrance of the coil—usually about 600°F.

TABLE 2 REACTOR-COIL DATA

Table tube length	200 ft
Tube ID	$\frac{1}{16}$ in.
Tube OD (outer diameter)	$\frac{1}{8}$ in.
Tube material	316 stainless steel, seamless
No. of individually controlled heating zones	9
Tube length per zone	22.2 ft
Helix dimensions	1-ft diameter $\times$ $2\frac{1}{2}$ ft high
Electrical resistance per 22.2-ft tube section	1 Ω
Transformer output	
Zones 1–6	35 V, 35 A
Zones 7–9	40 V, 40 A
Maximum power requirement for transformers (total)	12 kW
Maximum operating temperature	1600°F
Maximum design pressure	1000 psi
Temperature controller type	Weathermeasure, TRA-1, Triac-Triggered SCR gate

This is sufficiently high to inhibit steam condensation at the highest pressures used in this study but low enough to inhibit any significant reaction of coal solid. When both gas and solids flow are begun, the final desired temperature profile is established in the reactor tube. In the various tests conducted, the temperature either increases along the coil in the direction of gas–solids flow or was maintained at a constant value. For increasing temperatures, the temperature-distance characteristic along the coil corresponded to a linear relationship between the temperature and the gas–solids residence time in the coil of about 50°F/sec. In isothermal tests, gas–solids residence times 5–14 sec were employed.

The hot gas–solids mixture existing from the bottom of the reactor coil passes through an initial quench system that rapidly reduces its temperature to approximately 600°F to inhibit further reaction. At this point in the system, a lower temperature is avoided to prevent steam from condensing on the solids. The partially cooled mixture then proceeds through one of three solids filters, which retains the solids but permits gas flow. The gas continues through a condensor that removes water and oils and then passes through a gas-sampling panel, which is used intermittently to obtain gas samples for mass spectrographic analysis. In some of the tests conducted, the product gas was also continuously monitored with a Beckman Model 400 hydrocarbon analyzer to measure the total concentration of carbon in hydrocarbon species.

The data of primary interest in a given test corresponded to steady-state operation. Since a certain amount of time is required to achieve such operation, certain facilities are incorporated into the system to permit the collection of solid residues corresponding to steady-stage operation but not contaminated with residues resulting from unsteady-state operation. To accomplish this, the product solids could be collected in any of the three solids filters, depending on the position of a multiple-exit hot valve (valve V2 in Figure 1). During unsteady-state operation, when the desired gas and solids flows and the temperature profile in the reactor coil were being established, the product gas and solids flows were directed through solids filter A. When steady-state conditions were established, the product gas and solids were directed through solids filter B, which then accumulated a solids residue for analysis. Before the end of some tests, a direct determination of the solids inventory in the reactor coil was made to estimate the average solids residence times. This was accomplished by simultaneously closing valve V1 at the top of the coil, stopping the screw feeder, and diverting the product gas and solids flow through solids filter C. Valve V1 is a hot valve fitted with a solids filter that stops solids flow but permits gas flow when in a closed position. After these simultaneous operations, the solids inventory in the reactor coil is accumulated in solids filter C. Average solids residence times computed from chemical analyses and weight measurements of these solids generally corresponded very closely to calculated gas residence times, indicating negligible gas–solids slippage in the reactor coil.

Data Analysis

The experimental system employed in this study is an integral system in the sense that the gas, liquid, and solids conversion determined by analyses of the reactor-coil exit streams are the result of chemical interactions occurring along the length of the coil under systematically varying environmental conditions. With this type of system, the information required for proper kinetic characterization includes definitions of the conversions and local environmental conditions along the entire length of the coil, not only at the exit. Although this information could not be obtained in a single experimental test, a good approximation was achieved by series of properly designed tests. The basis for design of such test series depended on the fact that the gas and solids were essentially in plug flow through the coil, and slippage of the solids relative to the gas flow was negligible, since under these conditions both gas and solids conversions could be expressed solely as a function of pressure, initial gas:solids feed ratio, temperature, and temperature–time history.

Individual tests in isothermal test series were conducted at the same temperature, pressure, feed gas:solid ratio, and feed gas composition, varying only total feed gas and solids flow rates to obtain results as a function of residence time. Individual tests in test series conducted at constant heat-up rate conditions were designed to obtain results as a function of final temperature at the same pressure, feed gas:solids ratio, feed-gas composition, initial temperature, and heat-up rate. This was accomplished by varying the feed-gas flow rate and temperature profile in individual tests according to the following expressions:

$$G_0 = \frac{\pi d^2 P L \alpha}{2R(T_f^2 - T_0^2)} \quad [1]$$

and

$$T = \left[\frac{T_0^2 + (T_f^2 - T_0^2)z}{L}\right]^{1/2} \quad [2]$$

where

z = length at intermediate point along the reactor coil
L = total length of reactor coil
T_0 = temperature at entrance of reactor coil
T_f = temperature at ractor-coil exit (final temperature)
T = temperature at intermediate point z along the reactor coil
G_0 = feed-gas flow rate (mol/time)
R = gas constant
d = reactor-tube diameter
α = gas–solids heat-up rate
P = pressure

With this approach, and in the absence of catalytic reactor wall effects, yields obtained in individual tests conducted at a constant heat-up rate at various final temperatures could be interpreted as approximating yields occurring along the length of the reactor coil in a single test conducted at the maximum final temperature employed.

The question of reactor-wall-catalyzed reactions was investigated in a series of preliminary tests with simulated feed gases in the absence of coal solids. The results of these tests indicated that the only reaction of significance that occurred in the presence of typical concentrations of the major gas species was the water-gas shift reaction, which was initiated at approximately 1200°F.

RESULTS

Feed-gas compositions used in individual test series are given in Table 3 along with a definition of notation to distinguish primary results obtained in these test series, as illustrated in Figures 2-12. This notation is also applicable to Figures 13-14, 15 and 17. In the presentation of experimental results, various species and species groups that evolved during gasification have been categorized as follows:

- Carbon monoxide, carbon dioxide, hydrogen, methane, ethane, and benzene. Determined by mass-spectrographic analysis of dry product gases.
- Water. Computed as the difference between total oxygen gasified and oxygen present in carbon monoxide and carbon dioxide in the product gas.
- Unknown gaseous hydrocarbon. Computed as the difference between total carbon in gaseous hydrocarbon species, as determined by a hydrocarbon analyzer, and carbon present in methane, ethane, and benzene in the product gas.
- "Heavy hydrocarbon." Combined as the difference between total carbon gasified and carbon present in carbon monoxide, methane, and ethane.

The basis for this classification of species and species groups is related to the accuracies of analytical measurements in this study. Such measurements were limited by the fact that the concentrations of reaction products in the dry product gas in all tests conducted were less than 5% by volume (CO or CO_2, 0-1.5%; CH_4, 0-3%; C_2H_6, 0-0.8%; C_6H_6, 0-0.3%). These conditions were employed so

TABLE 3 FEED-GAS COMPOSITIONS

Notation in Figs. 2-15, 17	Feed Gas Pressure (atm)			Temperature Profile	
	H_2	He	Total	Isothermal	Constant Heat-up Rate
●	0	35	35	–	X
△	18	0	18	–	X
▲	18	17	35	–	X
□	35	0	35	–	X
▽	52	0	52	–	X
●	0	35	35	X	–
⊠	18	0	18	X	–
▲	18	17	35	X	–
⊠	35	0	35	X	–

that, in individual tests, the partial pressures of feed-gas components were essentially constant throughout the length of the reactor coil, which facilitates quantitative kinetic analyses.

The direct measurement of water yield in condensed liquid products was not usually accurate because of the relatively small amounts obtained and the uncertainty of the quantity of this species that was not condensed in the knockout pot. Computed values for the yield of this species are likely to be somewhat greater than actual yields because of the likelihood that some oxygen could be combined in oils and tars. This error, however, is probably small. The measurement of concentrations of gaseous hydrocarbon species, other than methane and ethane, and to a lesser degree benzene, by mass-spectrographic analysis was difficult for the test conditions used because of the small molecular concentrations of individual species. Although measurements of benzene concentrations in the gas were sufficiently accurate to be meaningful, the interpretation of this concentration in terms of total benzene yields is questionable because some of the benzene formed in the reactor coil may have condensed in the knockout pot but later may have vaporized when the liquid products warmed to ambient temperatures prior to containment.

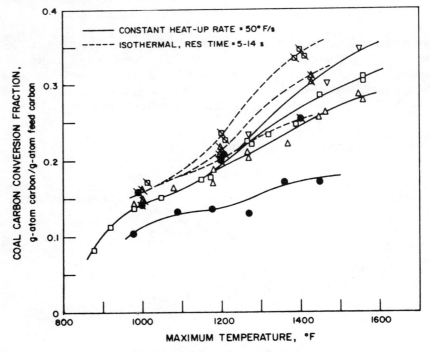

Figure 2 Coal-carbon conversion.

The unknown-gaseous-hydrocarbon group probably consists primarily of lighter aliphatic species such as ethylene, propane, propylene, butane, and butylene. The heavy-hydrocarbon group consists of the potentially condensible tars and oils, including benzene, and the unknown-gaseous-hydrocarbon species. In a few of the tests conducted, sufficient condensed hydrocarbon liquids were recovered to make direct experimental evaluations of total carbon balances. The fact that these balances showed better than 98% recovery suggests that, in the majority of tests in which insufficient liquids were recovered from quantitative analysis, the computed difference between the carbon in the heavy-hydrocarbon group and the carbon in the unknown-gaseous-hydrocarbon group is probably quite representative of the carbon in the actual condensed hydrocarbons, when benzene yields are negligible. The results given in Figures 1-12 exhibit the following major trends:

Evolution of Major Coal Components: Carbon, Oxygen, and Hydrogen (Figures 2-4)

The evolution of total carbon from the coal solids generally increases with increasing temperature and hydrogen partial pressure, with conversions obtained

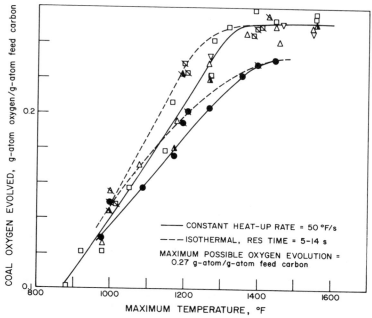

Figure 3   Coal-oxygen conversion.

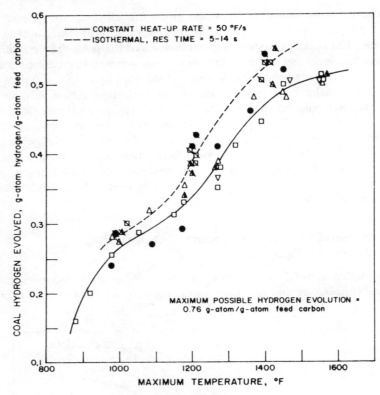

Figure 4 Coal-hydrogen conversion.

for isothermal operation being greater than for operation at constant heat-up rate. Total oxygen evolution from coal solids also increases with increasing temperature, and although conversions obtained in hydrogen are greater than in helium, no significant effect of hydrogen partial pressure on conversion is apparent in the range 18-52 atm. As with total carbon conversion, total oxygen evolution is also greater under isothermal operation. Total evolution of hydrogen from the coal solids increases with increasing temperature and is greater in isothermal tests but is *not* a significant function of hydrogen partial pressure. This is a particularly important result, indicating that hydrogen evolution is primarily a thermally activated phenomenon dependent only on time-temperature history.

It is also of significance that the results shown in Figures 2-4 for operation with a hydrogen-helium mixture (hydrogen partial pressure = 18 atm) are essentially identical to results obtained with pure hydrogen at a total pressure of 18 atm. This similarity is also apparent in yields of gasified products and suggests that hydrogen partial pressure, and not total pressure, is the main parameter affecting kinetic behavior during the initial gasification stages of Montana lignite. A somewhat different effect has been observed in an investigation with bituminous coal by Anthony [1], where it was found that initial gasification yields tended to increase with increasing hydrogen partial pressure but to decrease with increasing total pressure. This effect was explained as being due to increased diffusion resistance within the coal structure. This was not observed in this previous study with lignite, which does not become plastic during devolatilization.

Oxygen-containing Product Species (Figures 5 and 6)

For tests conducted at a constant heat-up rate, total carbon dioxide evolution is completed below 1000°F (Figure 5). The amount evolved (ca. 0.029 g-mol/g-atom feed carbon) is essentially the same in hydrogen and in helium and is in-

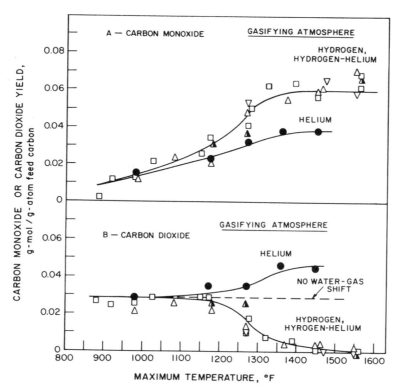

Figure 5 Carbon monoxide and carbon dioxide yields.

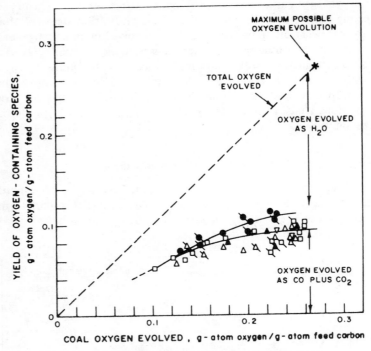

Figure 6   Adjusted evolved oxygen distribution.

dependent of hydrogen partial pressure. This amount is probably reflective of the concentration of carboxyl functional groups in the raw lignite. As temperature increases above 1000°F, the carbon dioxide yield remains substantially constant up to about 1200°F, then decreases with further increases in temperature for tests conducted in hydrogen, but increases with further increases in temperature for tests conducted in helium. These variations above 1200°F are probably due to the water-gas shift reaction, as suggested by results of tests conducted in the absence of coal solids. The dashed line shown in Figure 5b represents the assumed carbon dioxide yield for the case in which no water-gas shift occurs and was used as a basis for adjusting the total oxygen distribution in directly evolved species, as illustrated in Figure 6.

Hydrocarbon Product Species (Figures 7 to 12)

Methane-plus-ethane yields are highly dependent on hydrogen partial pressure (Figure 7). For tests conducted at a constant heat-up rate in hydrogen, methane-

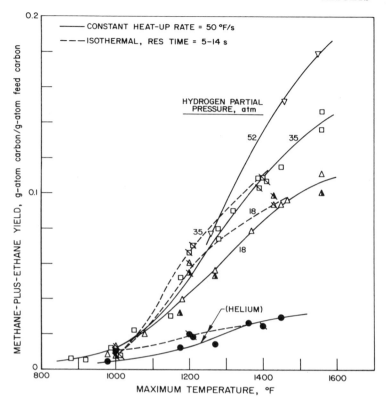

Figure 7 Methane-plus-ethane yields.

plus-ethane formation is slight below 1000°F (ca. 0.01 g-atom carbon/g-atom feed carbon). With further increases in temperature, yields in helium increase slightly, leveling off at a value of about 0.03 g-atom carbon/g-atom feed carbon above about 1300°F; in hydrogen, dramatic increases in methane-plus-ethane yields occur with increasing temperature. At about 1000-1200°F this increase is about the same for hydrogen partial pressures of 18-52 atm; above 1200°F, methane-plus-ethane yields increase with increasing hydrogen partial pressure, and yields at all pressures tend to suggest leveling off at higher temperatures. Reasonable extrapolation of the curves shown would indicate little increases in yields above about 1700°F for the reaction times employed.

Methane-plus-ethane yields obtained in isothermal tests are essentially independent of gas–solids residence times ranging 5-14 sec and are the same as yields obtained in constant heat-up rate tests at 1000°F and 1400°F. This is true for tests conducted in hydrogen and in helium. At 1200°F yields obtained in

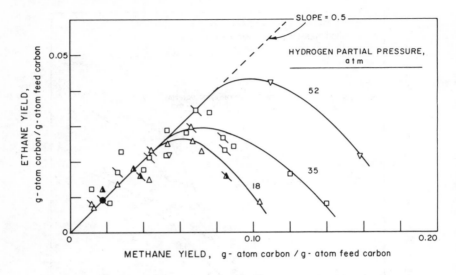

Figure 8 Relationship between methane and ethane yields.

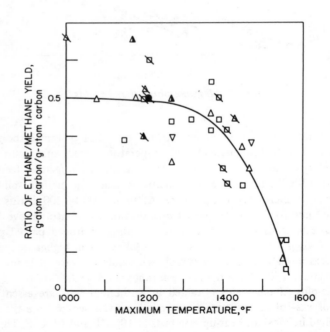

Figure 9 Effect of temperature on ethane: methane ratio.

isothermal tests are somewhat greater than those obtained in constant heat-up rate tests.

The sum of methane plus ethane has been referred to in the preceding discussion instead of the yields of each species individually because of an apparent stoichiometric relationship between the formation rates of each species. One such indication is illustrated in Figure 8, which shows that ethane yields are approximately directly proportional to methane yields up to values of methane yields of about 0.06 g-atom carbon/g-atom feed carbon. At higher methane yields, ethane yields tend to approach a maximum and then decrease with further increases in methane yields, with the maximum increasing with increasing hydrogen pressure. This behavior can be explained by assuming that, at all temperature levels, ethane is formed in direct proportion to methane, but that, at sufficiently high temperatures (above ca. 1300–1400°F), ethane converts to

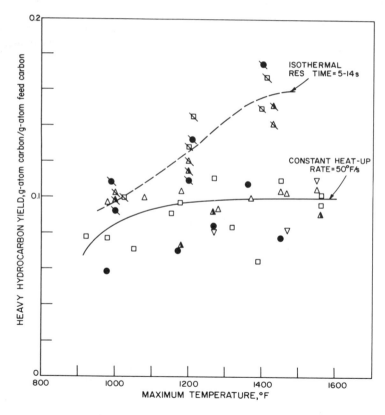

Figure 10 Heavy-hydrocarbon yields.

methane in the presence of hydrogen. The increasing maximum ethane yields with increasing hydrogen pressure can be explained by noting that these maxima occur at about the same temperature. This is demonstrated in Figure 9, which shows that the ratio of ethane:methane yield is apparently a function only of temperature and not pressure. This evidence suggests that hydrogen attack on lignite does not result in formation only of methane at any level of temperature, but rather results in a formation of both ethane and methane in a fixed ratio. Examination of data available in the literature [5, 6] suggests that this ratio tends to decrease with increasing coal rank.

The heavy-hydrocarbon yields shown in Figure 10 are substantially constant above 1000°F for tests conducted at constant heat-up rates, and yields obtained in hydrogen and in helium are generally similar. This species group—consisting of oils, tars, and light aliphatic gaseous species—apparently is formed below 1000°F, and although variations in the distribution of individual species within this group are likely at higher temperatures, there is apparently only limited transformation of species in this group to methane, ethane, or char, at least up to 1560°F. Heavy-hydrocarbon yields obtained in isothermal tests are significantly greater than those obtained in constant heat-up rate tests, particularly at 1400°F. This may occur primarily because, for isothermal operation, feed-coal solids heat up very rapidly to reactor temperature in the first few feet of the reactor coil. Assuming that most heavy-hydrocarbon formation occurs below

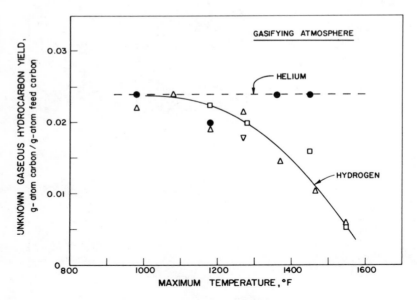

Figure 11   Unknown-hydrocarbon yields.

1000°F, increased heat-up rates through this range of temperature would tend to favor evolution of tars and oils, in competition with repolymerization in the solid phase. With this explanation, it is pertinent that increased isothermal temperature levels correspond to increased heat-up rates through the oil–tar formation range, being of the order of several thousand degrees per second at an isothermal temperature of 1400°F.

Figure 11 indicates that the unknown-gaseous-hydrocarbon yield in hydrogen decreases above about 1200°F. Although this decrease is not reasonably detectable in a corresponding decrease in the heavy-hydrocarbon yield, possibly because of data scatter above 1200°F, light aliphatic species can reasonably be expected to begin to convert to ethane and methane in the presence of hydrogen. This conversion does not occur in helium. Semiquantitative indications of benzene yields (Figure 12) suggest that the heavier components in the oil–tar

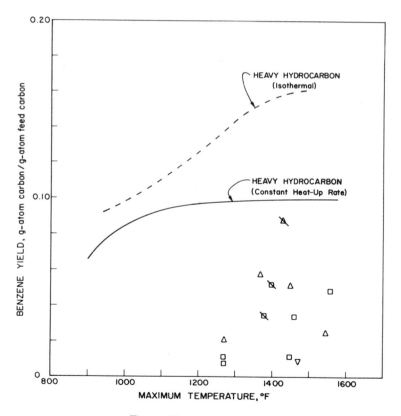

Figure 12 Benzene yields

fraction begin to convert to benzene at about 1300°F, with substantial conversions achieved by 1450°F. No benzene was detected in gas analyses for any test conducted below 1270°F with hydrogen, nor at any temperature for tests conducted in helium, suggesting that benzene is not a significant fraction of the oils and tars that initially evolve below 1000°F.

Relationship Between Equivalent Methane-Plus-Ethane Yield and Hydrogen Gasified

Figure 13 shows that, at any hydrogen pressure, "equivalent" methane-plus-ethane yields are directly proportional to the amount of hydrogen gasified. Equivalent methane-plus-ethane yields represent the difference between actual

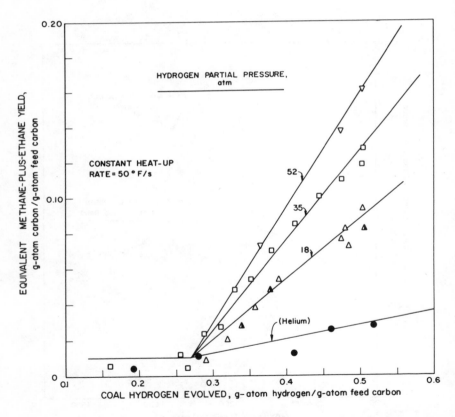

Figure 13 Stoichiometric relationship between equivalent methane-plus-ethane yield and coal–hydrogen conversion.

methane and ethane yields and an adjustment term obtained from Figure 11 at a corresponding temperature. The correction term is the difference between values of the unknown-gaseous-hydrocarbon yield indicated by the dashed line and the solid line. The basis for this correction is the assumption that the unknown-gaseous-hydrocarbon group consists of relatively low-molecular-weight aliphatic species (other than methane and ethane), which hydrogenate to form ethane and methane at increased temperatures in the presence of gaseous hydrogen. Because of a relationship is sought to characterize methane and ethane formation only as derived directly from the coal or coal char, the estimated amount of methane and ethane formed from gaseous interactions was subtracted from total methane-plus-ethane yields.

The relationship shown in Figure 13 is a very important one. It indicates that, at any hydrogen partial pressure, methane and ethane evolve directly in proportion to the total amount of hydrogen evolved from the coal, although the proportionality increases significantly with increasing pressure. It should be recalled that results given in Figure 4 show that total hydrogen evolution is not a function of hydrogen partial pressure and is essentially identical in hydrogen and helium. This combined evidence suggests then that the formation of active sites that catalyze methane and ethane formation in the presence of hydrogen is directly related to the process in which coal hydrogen is released since this latter process is independent of gaseous atmosphere and dependent only on time-temperature history. A model for quantitatively correlating equivalent methane and ethane formation rates from coal solids, based on the evidence discussed, is presented in the final section of this chapter.

CORRELATION OF "RAPID-RATE" METHANE AND ETHANE FORMATION

In consideration of the data obtained in this study with Montana lignite, the following model is proposed to describe the kinetics of methane and ethane formation during initial stages of gasification. During heat up of raw lignite (structure A), interactions within the coal initially occur below 1000°F and result primarily in the evolution of (1) carbon dioxide, probably resulting from gasification of all carboxyl functional groups, (2) some water and carbon monoxide, (3) some relatively low-molecular weight aliphatics, and (4) oils and tars. These reactions are essentially pyrolysis reactions that occur because of the breaking of certain of the weaker side-chain bonds as well as bonds connecting relatively large polyatomic molecules in the carbon matrix. This latter process results in the intermediate formation of large fragments, possibly free radicals, which then either (1) become stabilized because of hydrogen disproportionation or interaction with gaseous molecular hydrogen and evolve as oils and tars or

(2) polymerize to form an intermediate type of solid (structure B). The total amounts of materials other than oils and tars that gasify below 1000°F are essentially independent of gas atmosphere or heat-up rate, suggesting a stoichiometric relationship between the individual species formed and the functional groups present in the raw lignite. Total oil and tar formation is similar in hydrogen and in helium, and increases with increasing heat-up rate, which is apparently a result of the competition between stabilization and polymerization of intermediate free-radical fragments. At 1000–1300°F most of the remaining oxygen in the coal is evolved as carbon monoxide and water, and water formation is slightly greater in hydrogen than in helium. The gasification of hydrogen from the coal solids is relatively small at 1000–1200°F, due to the formation of water and some methane and ethane in the presence of gaseous hydrogen and both water and hydrogen in the presence of helium. Above 1200°F evolution of coal hydrogen begins to increase rapidly with increasing temperature, accompanied by a rapid increase in the formation of methane and ethane in the presence of gaseous hydrogen. The removal of oxygen at 1000–1300°F can be considered to correspond to the transition of the main carbon matrix from structure B to a second intermediate main structure (structure C). Structure C is considered to be comprised primarily of carbon and hydrogen and, with increases in temperature above about 1200°F, converts to a relatively stable "char" structure (structure E) through the evolution of hydrogen. During this transition, however, structure C initially converts to an active intermediate structure. Structure D, as hydrogen is evolved, and structure D then can either convert to the stable char structure, structure E, or interact with molecular hydrogen to form methane and ethane.

The following quantitative representation of the steps leading to methane and ethane formation assumes for simplicity that all oxygen is gasified prior to the formation of methane and ethane as a result of interactions of gaseous hydrogen with structure B. Although the experimental data indicate some overlap between the final stages of oxygen gasification and the initial stages of methane and ethane formation at 1200–1300°F, this assumption does not appreciably alter the quantitative evaluation of the parameters derived based on the model proposed.

The processes that lead to "rapid-rate" methane and ethane formation are assumed to occur according to the following overall reactions:

$$CH_x^0(s) \xrightarrow{k_0} CH_y^*(s) + (x-y) H(g) \qquad [1]$$

$$H_2(g) + CH_y^*(s) \begin{cases} \xrightarrow{k_1} (1-m) CH_y^0(s) \\ \xrightarrow{k_2} \left(\frac{m}{1+B}\right) CH_4(g) + \left[\frac{mB}{2(1+B)}\right] C_2H_6(g) + ymH(g) \end{cases} \qquad [2]$$

where

CH_x^0 = solid component resulting from interactions occurring during primary pyrolysis (structure C)
CH_y^* = intermediate solid active species (structure D)
H = hydrogen evolved from solids in Reactions 1 and 2
CH_y^0 = product coal char (structure E)
H_2 = gaseous molecular hydrogen
x = hydrogen:carbon atomic ratio in CH_x^0
y = hydrogen:carbon atomic ratio in CH_y^0 and CH_y^*
B = carbon ratio of ethane:methane formed in Reaction 2
m = fraction of carbon in CH_y^* converted to methane and ethane in Reaction 2
k_0, k_1, k_2 = first-order rate constants
s = solid
g = gas

Let

λ = fraction of feed carbon as CH_x^0 when conversion to structure C is complete (in g-atom carbon/g-atom feed carbon)
n_C^* = equivalent methane and ethane formed from coal at any time during gasification (in g-atom carbon/g-atom feed carbon)
n_C^0 = equivalent methane and ethane formed by pyrolysis reactions prior to the onset of Reaction 1 (in g-atom carbon/g-atom feed carbon)
f = fraction of CH_x^0 converted via Reaction 1 at any time during gasification
n_H^* = total coal hydrogen gasified at any time (in g-atom hydrogen/g-atom feed carbon)
n_H^0 = hydrogen gasified via pyrolysis reactions prior to the onset of Reaction 1 (in g-atom hydrogen/g-atom feed carbon)

Based on these definitions, it is possible to determine certain of the unknown stoichiometric parameters that characterize the model assumed, prior to consideration of the kinetics of Reaction 1. In these evaluations, it is assumed that the ratio k_2/k_1 is independent of temperature.

According to the preceding definitions,

$$f = \frac{n_C^* - n_C^0}{m\lambda} \qquad [3]$$

and

$$f = \frac{n_H^* - n_H^0}{(x - y + ym)\lambda} \qquad [4]$$

Combining Equations 3 and 4 and rearranging leads to

$$n_C^* = \frac{mn_H^*}{x - y + ym} + n_C^0 - n_H^0 \frac{m}{x - y + ym} \qquad [5]$$

Letting $S = [m/(x - y + ym)]$ and $I = [n_C^0 - n_H^0 m/(x - y + ym)]$, Equation 5 can be represented as

$$n_C^* = S n_H^* + I \qquad [6]$$

Thus where m is constant, a plot of n_C^* versus n_H^*, both experimental parameters, should yield a straight line with a slope equal to S and intercept at $n_H^* = 0$ of I. Figure 13 shows such a plot for data obtained in constant heat-up rate tests in helium and in hydrogen. Values of S increase with increasing pressure because of increasing values of m with increasing pressure. The common point of intersection of the lines drawn corresponds to values of $n_C^0 = 0.01$ g-atom carbon/g-atom feed carbon and $n_H^0 = 0.272$ g-atom hydrogen/g-atom feed carbon. Table 4 tabulates the values of S obtained from Figure 13 as a function of hydrogen partial pressure, P_{H_2}.

Now, let

Y_C = total carbon in partially gasified lignite (g-atom carbon/g-atom feed carbon)

Y_H = total hydrogen in partially gasified lignite (g-atom hydrogen/g-atom feed carbon)

$Z = (Y_H/Y_C)$ = hydrogen:carbon ratio in partially gasified lignite (g-atom hydrogen/g-atom carbon)

TABLE 4 VARIATION IN S WITH P_{H_2}

P_{H_2} (atm)	S (g-atom Carbon/g-atom Hydrogen)
0	0.084
18	0.352
35	0.514
52	0.649

From the stoichiometry defined in Reactions 1 and 2, Z is given by the expression

$$Z = \frac{x - f(x - y + ym)}{1 - fm} \qquad [7]$$

Solving for f in Equation 7 results in

$$f = \frac{x - Z}{x - y + ym - mZ} \qquad [8]$$

Equating the expression for f in Equation 8 to the expression for f in Equation 3 leads to

$$\frac{n_C^* - n_C^0}{\lambda m} = \frac{(x - Z)/m}{(x - y + ym)/m - Z} \qquad [9]$$

Rearranging Equation 9 and substituting S for the expression defined results in

$$(n_C^* - n_C^0) \frac{1}{S - Z} = \lambda(x - Z) \qquad [10]$$

Thus a plot of the term on the left-hand side of Equation 10 versus Z should result in a straight line with a slope equal to $-\lambda$ and an intercept at $Z = 0$ of λx. Figure 14 shows such a plot for constant heat-up rate tests conducted in hydrogen. Data obtained with helium were not included in this plot because of the scatter that results from small values of S, which magnify variations in the term $(n_C^* - n_C^0)$. In accordance with the model assumed, data obtained at hydrogen partial pressure of 18 atm, 35 atm, and 52 atm are reasonably correlated with a single straight line corresponding to a value of $\lambda = 0.83$ g-atom carbon/g-atom feed carbon and a value of $x = 0.578$ g-atom hydrogen/g-atom carbon. The value of λ was not determined by a least-squares fit of the data but was "forced" so that the amount of carbon initially present in the component CH_x^0 is equal to the total amount of carbon initially present in the raw lignite, less the total carbon evolved during pyrolysis due to formation of carbon monoxide, carbon dioxide, and the heavy-hydrocarbon species. Note that if the component CH_x^0 consisted of polycondensed aromatic units of hexagonally arranged carbon, with hydrogen present on lattice edges, then the value of $x = 0.578$ corresponds to an average ring number of 5.

The value of y is assumed to be 0.25 g-atom hydrogen/g-atom carbon, based on measurements made of hydrogen contents of Montana lignite chars gasified at elevated temperatures for extended times. With this assumption, values of m can

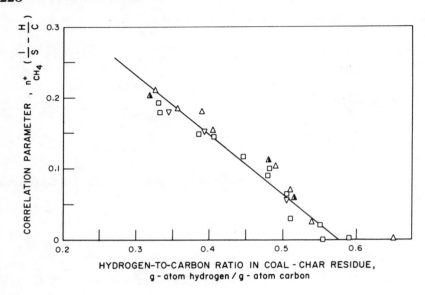

Figure 14 Relationship between stoichiometric correlation parameter and coal hydrogen: carbon ratio.

be computed for corresponding values of S obtained from Figure 13, according to the expression

$$m = \frac{S(x-y)}{1-Sy} \qquad [11]$$

In addition, values of $r = (k_2/k_1) = m/(1-m)$ can also be computed. Values of m and r are given in Table 5 as a function of hydrogen pressure for constant heat-up rate tests.

TABLE 5 VARIATION OF m AND r WITH P_{H_2}

P_{H_2} (atm)	m	r
0	0.028	0.029
18	0.127	0.145
35	0.194	0.240
52	0.254	0.341

The values of r given in Table 5 increases with increasing hydrogen partial pressure. Figure 15 shows, in fact, a linear relationship between r and P_{H_2}, which is represented by the expression

$$r = 0.03 + 0.00605 P_{H_2} \qquad [12]$$

Thus all the parameters necessary to quantitatively characterize the model assumed have been determined, except for parameters indicative of the kinetics of Reaction 1. After complete conversion of the reaction intermediate, CH_y^*, however, the maximum methane-plus-ethane yield is independent of the kinetics of Reaction 1 and can be expressed as a function of the hydrogen partial pressure by the following expression:

$$\text{Maximum methane-plus-ethane yield, g-atom carbon/g-atom feed carbon} = \frac{0.83(0.029 + 0.00587 P_{H_2})}{1 + 0.00587 P_{H_2}}$$

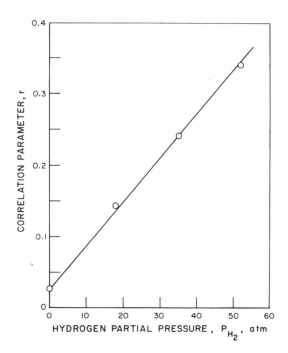

Figure 15 Effect of hydrogen partial pressure on kinetic parameter, r.

230 MONTANA LIGNITE

It is of interest that the empirical form of this expression is essentially the same as an expression proposed by Zahradnik and Grace [12] to relate total methane yield as a function of hydrogen partial pressure during the initial hydrogenation of coals.

The following assumptions were made, consistent with results obtained in both constant heat-up rate and isothermal tests, to describe the kinetics of Reaction 1: (1) CH_x^0 reacts according to Reaction 1 by a first-order process, but where there is a distribution of activation energies for the first-order rate constant, k_0; and (2) the distribution function of activation energies is a constant; that is, $f(E)dE$ = fraction of total carbon in which the activation energy E in the rate constant

$$k_0 = k_0^0 \exp\left(\frac{-E}{RT}\right)$$

is between E and $E + dE$

where

$f(E) = 0$ for $E < E_0$
$f(E) = C$ (constant) for $E_0 \leq E \leq E_1$
$f(E) = 0$ for $E > E_1$
k_0^0 = preexponential factor

Note that because

$$\int_{E_0}^{E_1} C\, dE = 1$$

then

$$C = \frac{1}{E_1 - E_0}$$

From these assumptions, the average conversion fraction of CH_x^0 can be expressed by the following relationship for any time–temperature history:

$$1 - f = \frac{1}{E_1 - E_0} \int_{E_0}^{E_1} \left\{ \exp\left[-k_0^0 \int_0^\theta \exp\left(\frac{-E}{RT}\right) d\theta\right] \right\} dE \qquad [14]$$

where

θ = time
T = absolute temperature
R = gas constant

For the specific case of constant heat-up rate, where $\alpha = (dT/d\theta)$, Equation 14 can be expressed as the following:

$$1 - f = \frac{1}{E_1 - E_0} \int_{E_0}^{E_1} \left\{ \exp\left[- \frac{k_0^0}{\alpha} \int_{T_0}^{T} \exp\left(\frac{-E}{RT}\right) dT \right] \right\} dE \quad [15]$$

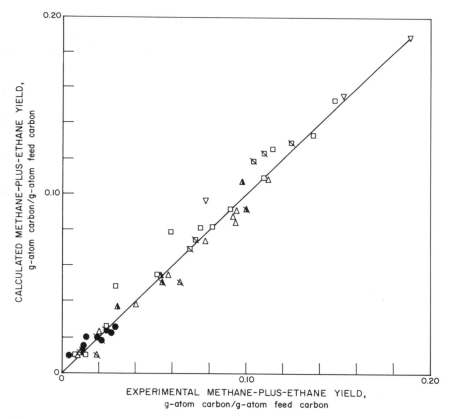

Figure 16 Comparison of calculated and experimental methane-plus-ethane yields.

232 MONTANA LIGNITE

For isothermal conditions, Equation 14 has the form

$$1 - f = \frac{1}{E_1 - E_0} \int_{E_0}^{E_1} \left\{ \exp\left[-k_0^0 \, \theta \, \exp\left(\frac{-E}{RT}\right)\right] \right\} dE \quad [16]$$

The best fit of our experimental data was obtained with the following values of E_0, E_1, and k_0^0:

$$E_0 = 79,500 \text{ cal/g-mol}$$

$$E_1 = 118,100 \text{ cal/g-mol}$$

$$k_0^0 = 7 \times 10^{20} \text{ sec}^{-1}$$

Experimental and calculated equivalent methane-plus-ethane yields are compared in Figure 16, where calculated yields were determined based on the parameters given in the preceding equations, using the appropriate correlation form for isothermal or constant-heat-up rate operation.

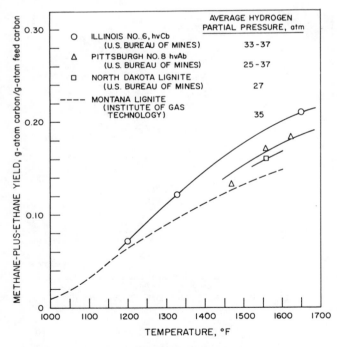

Figure 17 Comparison of methane-plus-ethane yields with different coals.

In summary, the correlations described in this chapter provide a basis for predicting methane and ethane yields as a function of temperature, hydrogen partial pressure, and time-temperature history during the gasification of Montana lignite. Although these correlations were developed based on data obtained in hydrogen, helium, and hydrogen-helium mixtures, the results of a previous thermogravimetric study conducted at the IGT [7, 10] showed that "rapid-rate" methane formation kinetics with air-pretreated bituminous coal are a function of hydrogen partial pressure even in gas mixtures containing other synthesis gas species, which may also be the case for the Montana lignite used in this current study. The generality of the parameters evaluated is, of course, not known because experimental results obtained only with Montana lignite were used in the development of the model. Although future studies are anticipated using other coals to evaluate this aspect, there is some evidence currently available from studies conducted at the U.S. Bureau of Mines, reported by Feldmann, Mima, et al. [4], in which results obtained for hydrogasification of raw bituminous coals in a 3-in.-ID transport reactor show a strong similarity to corresponding results obtained in this study with Montana lignite. Figure 17, for example,

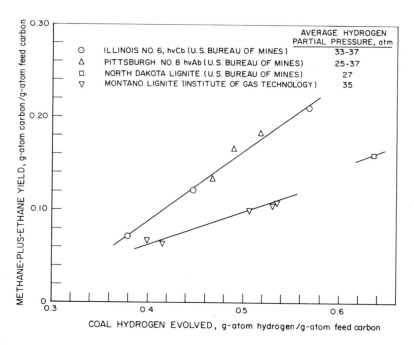

Figure 18 Comparison of methane-plus-ethane yields versus coal-hydrogen conversion characteristics for different coals.

compares methane and ethane yields obtained under isothermal temperature operation as a function of temperature for the two studies considered, at approximately corresponding average hydrogen partial pressures.[1] The results shown indicate that relatively minor adjustments in one or two of the parameters defined in the model proposed in this chapter would be required to fit the data obtained with the bituminous coals and the North Dakota lignite. Another data comparison is shown in Figure 18, which plots total methane and ethane yields versus total coal hydrogen evolution for these same coals. Although the model proposed predicts linearity in these relationships only when referring to "equivalent" methane-plus-ethane yields and only when total oil/tar yields do not vary, the results shown are nevertheless suggestive that equivalent methane-plus-ethane yields obtained with the bituminous coals are essentially proportional to coal hydrogen evolution, as was indicated in this study with Montana lignite.

REFERENCES

1. Anthony, D. G., "Rapid Devolatilization and Hydrogasification of Pulverized Coal," Sc.D. thesis, Massachusetts Institute of Technology, Cambridge, Mass., 1974.
2. Blackwood, J. D., and McCarthy, D. J., "The Mechanism of Hydrogenation of Coal to Methane," *Aust. J. Chem.*, 19, 797–813 (1966).
3. Feldmann, H. F., "Reaction Model for Bituminous Coal Hydrogenation in a Dilute Phase," paper presented at the 160th National Meeting of the American Chemical Society, Division of Petroleum Chemistry, Chicago, September 13–18, 1970.
4. Feldmann, H. F., Mima, J. A., and Yavorsky, P. M., "Pressurized Hydrogasification of Raw Coal in a Dilute-Phase Reactor," *Coal Gasification*, in *Advances in Chemistry Series*, No. 131, American Chemical Society, Washington, D.C., 1974.
5. Hiteshue, R. W., Friedman, S., and Madden, R., "Hydrogasification of Bituminous Coals, Lignite, Anthracite, and Char," *U.S. Bur. Mines, Rept. Invest.* (6125) (1962).
6. Hiteshue, R. W., Friedman, S., and Madden R., "Hydrogasification of a High-Volatile A Bituminous Coal," *U.S. Bur. Mines, Rept. Invest.* (6376) (1964).
7. Johnson, J. L., "Kinetics of Bituminous Coal Char Gasification with Gases Containing Steam and Hydrogen," *Coal Gasification*, in *Advances in Chemistry Series*, No. 131, American Chemical Society, Washington, D.C., 1974.

---

[1] The study at the U.S. Bureau of Mines was conducted under integral conditions, and the results presented correspond to operation with hydrogen-methane feed-gas mixtures. The average hydrogen partial pressure indicated in Figure 17 is the linear average of the feed and product gases.

8. Mosley, F., and Patterson, D., "The Rapid High-Temperature Hydrogenation of Coal Chars. Part 2: Hydrogen Pressures up to 1000 Atmospheres," *J. Inst. Fuel*, 38, 378–391 (1965).
9. Mosley, F., and Patterson, D., "The Rapid High-Temperature High-Pressure Hydrogenation of Bituminous Coal," *J. Inst. Fuel*, 40, 523–530 (1967).
10. Pyrcioch, E. J., Feldkirchner, H. L., Tsaros, C. L., Johnson, J. L., Bais, W. G., Lee, B. S., Schora, F. C., Huebler, J., and Linden, H. R., "Production of Pipeline Gas by Hydrogasification of Coal," *IGT Research Bulletin*, No. 39, Chicago, November 1972.
11. Zahradnik, R. L., and Glenn, R. A., "Direct Methanation of Coal," *Fuel*, 50, 77–90 (1971).
12. Zahradnik, R. L., and Grace R. J., "Chemistry and Physics of Entrained Coal Gasification," *Coal Gasification*, in *Advances in Chemistry Series*, No. 131, American Chemical Society, Washington, D.C., 1974.

4 Relationship Between Gasification Reactivities of Coal Char and Physical and Chemical Properties of Coal and Coal Char

INTRODUCTION

Various experimental investigations have studied, at elevated pressures, the gasification kinetics of coal chars in hydrogen and in gases containing steam and hydrogen. Most of these investigations, however, have been primarily concerned with the characterization of gasification rates as a function of environmental conditions such as temperature, pressure, and gas composition and have provided little systematic information concerning relationships between gasification reactivities and the physical and chemical properties of coal or coal char. Therefore, this study was initiated to evaluate possible relationships between gasification reactivities and simple compositional parameters for coal chars derived from a wide variety of coals and coal maceral concentrates. The internal structural changes that occur during the course of gasification of a few coal chars of varying rank were also explored. The gasification reactivities of individual coal chars were determined in hydrogen or in a 50:50 steam: hydrogen ratio at 35 atm, using a high-pressure thermobalance. Most tests were conducted at 1700°F, although, in a special series of tests designed to investigate the catalytic effects of exchangeable cations on lignite-char reactivities, temperatures were varied from 1400°F to 1700°F. The following results are discussed in this chapter.

- The relationship between the initial carbon content and the gasification reactivities determined in the hydrogen at 1700°F, for coal chars derived

Adapted from a paper presented at the 170th Meeting of the ACS/DFC, Chicago, August 24-29, 1975, Fuel Chemistry Division, ACS, Preprints 20 (4): 85-101 (1975), with permission.

from 36 coals and maceral concentrates ranging in rank from anthracite to lignite.
- The effect of exchangeable cation concentrations (sodium and calcium) on gasification reactivities of lignites in hydrogen at 1700°F and in steam-hydrogen mixtures at 1400-1700°F.
- The surface area and pore volume variations that occur during gasification in hydrogen and in steam-hydrogen mixtures at 1700°F of coal chars derived from anthracite, metallurgical coking coal, high-volatility A bituminous coal, subbituminous A coal, and lignite.

EXPERIMENTAL PROCEDURE

The high-pressure thermobalance used in this work to obtain gasification reactivity factors has been described previously [4]. The main feature of this apparatus is that the weight of a small, fixed-bed sample of coal char (0.5-1 g) contained in a wire-mesh basket can be continuously measured as it undergoes gasification in a desired gaseous environment at constant temperature and pressure. In all tests conducted, -20 + 40 U.S. sieve-sized particles were used, and gas-flow rates in the reactor were maintained at sufficiently high values to result in negligible gas conversion. Under these conditions, coal-char gasification could be considered to occur under constant known environmental conditions. Coal chars were produced by initially exposing the raw coals to nitrogen at 1 atm for 60 min, at the same temperature to be used during subsequent gasification in hydrogen or in steam-hydrogen mixtures. The weight loss versus time characteristics obtained during gasification in individual tests were then used as a basis for computing gasification reactivity factors, using a procedure described in the following paragraphs.

Certain of the solid feeds and residues were analyzed for internal surface area, pore volume, and true density. Surface areas were computed from adsorption isotherms obtained with a Model 2100 Orr surface area-pore volume analyzer manufactured by the Micromeretics Corporation, which was also used to obtain true densities in helium. Adsorption isotherms obtained in nitrogen at 77°K were interpreted with the BET equation to compute surface area, and isotherms obtained in carbon dioxide at 298°K were interpreted with the Dubinin-Polanyi equation as modified by Kaganer [5] to compute surface area. In general, surface areas computed from nitrogen and carbon dioxide adsorption isotherms were not in agreement, and values obtained in carbon dioxide were considered to be most reflective of equivalent internal surface area, which is consistent with the findings of other investigators [2, 3, 6, 7, 12]. Apparently, the penetration of nitrogen into the microporous structure of coals or carbonized coal chars is severely limited by slow, activated diffusion processes at 77°K, leading to very low apparent surface areas; on the other hand, for partially gasified coal chars having more open microporous structures, capillary condensation of nitrogen can lead to unreasonably high apparent surface areas [1, 8]. Adsorption isotherms obtained with carbon dioxide at the higher temperature of 298°K facilitate activated diffusion into microporous structures, and capillary condensation is inhibited by the lower relative pressures employed (0.003-0.02). Although there is some question as to whether carbon dioxide adsorption isotherms should be interpreted in terms of micropore volume [8-11] rather than micropore surface area [2, 3, 6, 7, 12], the distinction is not of importance in empirical correlations with gasification kinetic parameters. This is because the calculation methods used to compute numerical values of micropore volume and micropore surface area are virtually identical, differing only in the numerical constants used. Thus reported values of micropore volumes can be converted to corresponding values of micropore surface area by a fixed constant. In this study we have chosen to compute surface-area values, favoring the argument that carbon dioxide adsorption on a carbon surface should be restricted to a monolayer thickness, as a result of the quadruple interaction of the carbon dioxide molecules with the π bonds of the carbon surface [2, 7].

An Aminco mercury intrusion porosimeter capable of a hydrostatic pressure of 15,000 psi was used to obtain pore volume distributions for pores having diameters greater than about 120 Å. Pore volume distributions for pores of about 12-300 Å were obtained from adsorption isotherms obtained in nitrogen at 77°K at relative pressures up to about 0.93. Good agreement was obtained with these two methods in the overlap region of 180-300 Å, similar to results reported by Gan, Nandi, and Walker [3].

DEFINITION OF RELATIVE REACTIVITY FACTOR, f_L

Weight loss versus time characteristics obtained in individual thermobalance tests were interpreted to obtain relative reactivity factors for the coal chars used, based on a quantitative model developed previously at the IGT to describe the gasification kinetics of bituminous coal chars as a function of temperature, pressure, gas composition, pretreatment temperature, and carbon conversion [4]. The essential features of this model are described in the paragraphs that follow.

Coal-char gasification in gases containing steam and hydrogen are assumed to occur via three main reactions:

$$C + H_2O \underset{}{\overset{k_1}{\rightleftharpoons}} CO + H_2 \qquad [1]$$

$$C + 2H_2 \underset{}{\overset{k_2}{\rightleftharpoons}} CH_4 \qquad [2]$$

$$2C + H_2 + H_2O \underset{}{\overset{k_3}{\rightleftharpoons}} CO + CH_4 \qquad [3]$$

where k_1, k_2, and k_3 are rate constants for the three reactions shown and are quantitatively defined in the model as a function of temperature, pressure, and gas composition (accounting for CO, H_2, H_2O, and CH_4). The differential coal-char conversion rate is expressed by the relationship

$$\frac{dX}{d\theta} = f_L k_T (1 - X)^{2/3} \exp(-\alpha X^2) \qquad [1]$$

where

X = base-carbon conversion fraction
θ = time
f_L = relative reactivity factor for coal-char gasification, which depends on the particular coal char and on the pretreatment temperature used in preparing the coal char
$k_T = k_1 + k_2 + k_3$
α = kinetic parameter defined as a function of pressure and gas composition

"Base carbon," referred to in the preceding definition of X, means the nonvolatile carbon in raw coal that remains after standard devolatilization. Base-carbon conversion fractions can be estimated from weight loss–time curves obtained in thermobalance tests, using the expression

$$X = \frac{(W/W_0 - VM)}{1 - VM - A} \qquad [2]$$

where

W/W_0 = total weight-loss fraction referred to original coal
VM = weight loss fraction during initial devolatilization in nitrogen (approximately equal to standard volatile matter at elevated temperature)
A = ash mass fraction in feed coal

The relative reactivity factor, f_L, depends on the pretreatment temperature according to the expression:

$$f_L = f_0 \exp(8467) \left(\frac{1}{T_p} - \frac{1}{T}\right) \qquad [3]$$

DEFINITION OF RELATIVE REACTIVITY FACTOR, f_L

where

f_0 = reactivity factor dependent only on the inherent nature of the coal char
T_p = pretreatment temperature (in °R)
T = gasification temperature (in °R)

Equation 3 is only applicable for $T_p > T$; for $T_p \leq T$, then $f_L = f_0$.

At constant environmental conditions, Equation 1 can be integrated to yield

$$M(X) = \int_0^X \frac{\exp(\alpha X^2)}{(1-X)^{2/3}} dX = f_L k_T \theta \quad [4]$$

Based on Equation 4, a plot of $M(X)$ versus θ should yield a straight line having a slope equal to the term $f_L k_T$. Values of $M(X)$ can be computed from experimental thermobalance data, using Equation 2 to obtain values of X and using the defined value of α to evaluate the integral in Equation 4. Note that, for tests conducted in pure hydrogen, the value of α is 0.97; with this value, the term $(1-X)^{2/3} \exp(0.97 X^2)$ is approximately equal to $(1-X)$, over nearly a complete range of X. For this case, $M(X) = -\ln(1-X)$ and values of the specific gasification rate, $(dX/d\theta)/(1-X)$, are constant and equal to $f_L k_T$.

Figure 1 shows two types of behavior noted in this study in experimental

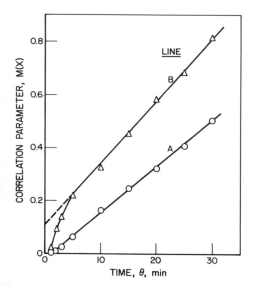

Figure 1 Types of correlation characteristics.

plots of $M(X)$ versus θ. Line A is typical of the characteristics obtained with the majority of coal chars tested, with linearity exhibited over the complete range of base carbon conversion. The line shown does not extrapolate to the origin because char samples initially exposed to the gasifying environment in the thermobalance require 1-2 min to heat up to reactor temperature. A characteristic of the type shown for case B was obtained with some coal chars, usually low-rank materials, indicating an initial period of transient reactivity, which decreased during the first 5-10 min and remained constant thereafter. For coal chars exhibiting this type of behavior, only the linear portion of the curve corresponding to constant reactivity was used to evaluate experimental values of $f_L k_T$.

Values of the reactivity factor, f_L, were then obtained by dividing the term $f_L k_T$ by the value of k_T defined in the model [4] for the reaction conditions used.

RESULTS

Correlation of Reactivity Factors with Carbon Content in Raw Coals

The reactivity factors for coal chars derived from 36 coals and coal maceral concentrates were determined in hydrogen at 1700°F and 35 atms. The distribution of coals used with respect to rank and lithotype is described in Table 1. In this study a variety of correlations were evaluated in attempting to quantitatively relate these reactivity factors with simple, compositional parameters included in ultimate, proximate, and petrographic analyses of the raw coals. Maximum success, however, was achieved with one of the simplest correlations considered—a relationship between reactivity factors and initial carbon contents. This correlation is illustrated in Figure 2, where the line drawn corresponds to

TABLE 1 NUMBERS OF EACH COAL TYPE USED IN CORRELATION

Coal Rank	Whole	Vitrain	Fusian	Total
Lignite	1	1	1	3
Subbituminous C	2	1	—	3
Subbituminous B	1	2	2	5
Subbituminous A	3	2	1	6
High-Volatile C Bituminous	1	2	1	4
High-Volatile B Bituminous	1	1	4	6
High-Volatile A Bituminous	4	2	—	6
Low-Volatile Bituminous	2	—	—	2
Anthracite	1	—	—	1
	16	11	9	36

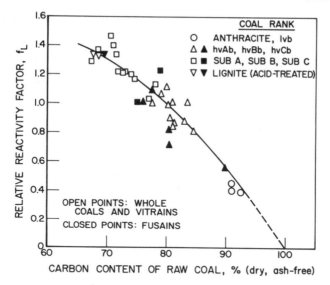

Figure 2 Correlation of reactivity factors.

the expression:

$$f_L = 6.2Y(1 - Y) \qquad [5]$$

where

f_L = relative reactivity factor of coal char

Y = concentration of carbon in raw coal (dry, ash-free) (in g/g coal)

For reasons discussed in the following paragraphs, it should be emphasized that, for lignite coal chars, the correlation shown in Figure 2 is applicable only when the raw lignite is initially treated in acid to remove exchangeable cations.

The standard deviation of experimental reactivity factors shown in Figure 2 and of reactivity factors calculated from Equation 5 is about 0.1, which is equivalent to the reproducibility of experimentally determined reactivity factors. Interestingly, the correlation proposed does not uniquely distinguish between maceral types.

Effects of Exchangeable Cation Concentration of Lignite Char Reactivity

If reactivity factors determined for coal chars derived from untreated raw lignites were included in Figure 2, a considerable amount of scatter would be apparent above the correlation line at low carbon concentrations. One phase of this study, however, showed that the reactivities of lignite chars obtained from lignites initially treated in HCl or HCl-HF acid were generally significantly less than the corresponding reactivities exhibited by lignite chars derived from untreated lignites. This was not observed with several bituminous and subbituminous coal chars. This behavior apparently resulted from a catalytic effect of exchangeable cations inherently present in raw lignites in carboxyl functional groups, which can be removed in acid by the following type of reaction:

$$-COONa + H^+ = -COOH + Na^+ \qquad [4]$$

With this explanation, one can reasonably expect that this catalytic effect would predominate in lignites and would decrease rapidly with increasing coal rank, corresponding to a rapid decrease in the amount of coal oxygen combined in carboxyl functional groups.

A series of tests were conducted to obtain a quantitative measure of the effects of exchangeable cation concentration (sodium and calcium) on char reactivity factors for gasification in hydrogen and in steam-hydrogen mixtures. These tests were conducted with lignite chars derived from raw lignites, with the lignite chars derived from raw lignites initially demineralized in hydrochloric acid to remove exchangeable cations and with the lignite chars derived from raw lignites initially demineralized in hydrochloric acid to which various amounts of calcium or sodium were then added by cation exchange in sodium acetate or calcium acetate solutions. Results of one series of tests corresponding to gasification in hydrogen at 1700°F are shown in Figure 3. The results in Figure 3 were correlated with the expression

$$\frac{f_L}{f_L^0} = 1 + 54.7 Y_{Na} + 14 Y_{Ca} \qquad [6]$$

where

f_L = reactivity factor of lignite to which sodium or calcium was added

f_L^0 = reactivity factor of acid-treated lignite (Y_{Na}, Y_{Ca} = 0)

Y_{Na}, Y_{Ca} = concentration of exchangeable sodium or calcium in lignite before devolatilization in nitrogen, g/g fixed carbon.

Although the correlation given in Equation 6 was developed from data obtained with prepared lignites that did not contain both calcium and sodium at the same time, it does apply reasonably well to untreated lignites containing, in some cases, both calcium and sodium. This is demonstrated in Table 2.

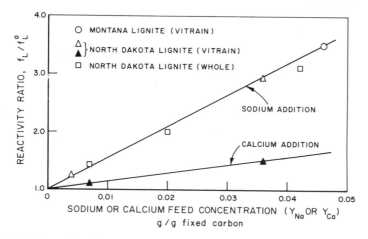

Figure 3 Effect of cation concentration on reactivity of lignite char in hydrogen at 1700°F and 35 atms.

A test series was also conducted to determine effects of exchangeable calcium and sodium concentrations on the reactivities on a Montana lignite char in steam–hydrogen mixtures at 1400–1700°F. Results obtained are illustrated in Figure 4, which plots values of the kinetic term, $f_L k_T$, as a function of temperature and cation concentration. Although these results have not yet been quantitatively correlated, they apparently show that sodium and calcium significantly enhance gasification in steam–hydrogen mixtures, even more so than for gasification in hydrogen alone. Figure 4 shows that the effect of calcium concentration on reactivities is substantially the same as the effect of sodium concentration at corresponding conditions (contrary to the behavior obtained in pure hydrogen),

TABLE 2 COMPARISON OF CALCULATED AND EXPERIMENTAL REACTIVITY RATIOS

Lignite	Y_{Ca}	Y_{Na}	f_L/f_L^0	
	g/g fixed carbon		Calculated	Experimental
Savage Mine, Montana (whole)	0.043	0.000	2.1	1.7
Savage Mine, Montana (vitrain)	0.019	0.004	1.5	1.7
Glenharold Mine, N. Dakota (whole)	0.031	0.009	1.9	2.0
Glenharold Mine, N. Dakota (vitrain)	0.019	0.002	1.3	1.2

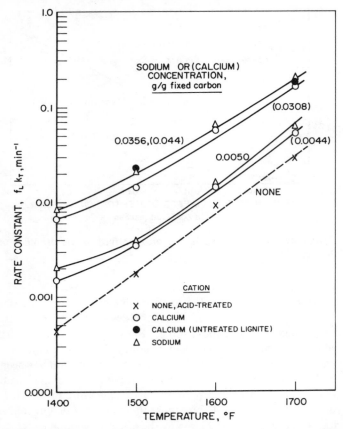

Figure 4 Effect of cation concentration on reactivity of Montana lignite char in steam-hydrogen mixtures at 1400–1700°F and 35 atms.

and that relative catalytic effects tend to decrease with increasing gasification temperature.

A significant additional result of the test series conducted with steam-hydrogen mixtures was that the reactivity of acid-treated Montana lignite (Y_{Ca}, Y_{Na} = 0) remains constant for gasification in steam-hydrogen mixtures over a temperature range of 1400–1700°F. This is shown in Figure 5, which plots experimental values of $f_L k_T$ versus values of k_T calculated from correlations developed to describe bituminous coal-char-gasification kinetics [4]. The line drawn corresponds to a constant value of f_L = 1.3, which is about the same value obtained for gasification in pure hydrogen at 1700°F.

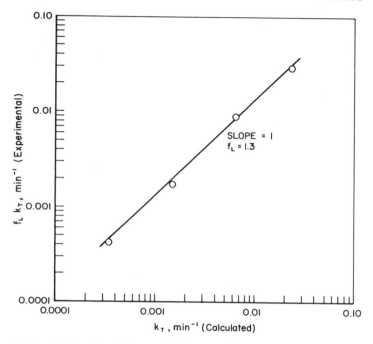

Figure 5 Reactivity of acid-treated Montana lignite char in steam-hydrogen mixtures at 1400–1700°F and 35 atms.

Variations in Internal Char Surface Areas During Gasification

The variations in internal surface areas were measured for several chars at different stages of gasification in hydrogen or steam-hydrogen mixtures. The compositions of the coals from which these chars were prepared are given in Table 3. Figure 6 shows the variations in surface area measured in carbon dioxide with different base carbon conversion fractions for a series of tests conducted with Montana lignite chars. Figure 6 shows that the internal surface area of the Montana lignite char tends to remain constant over a major range of base carbon conversion fractions and is essentially independent of char pretreatment or gasification conditions. The apparent surface areas of carbonized chars ($X = 0$) are lower than the nominal value of partially gasified chars. This difference may reflect that, even with carbon dioxide, penetration into the micropore structure is somewhat inhibited before the structure is opened up by partial gasification.

Figure 7 shows variations in apparent surface area measured in nitrogen for

TABLE 3 COMPOSITIONS OF COALS USED IN PHYSICAL-PROPERTY STUDIES

	Anthracite	Pittsburgh No. 8 (Ireland Mine) (Air Pretreated)	Brazilian Carvao Metallurgical	Rosebud, Colstrip Mine	Savage Mine (Montana)
Coal identification					
Rank	Anthracite	hvAb		Subbituminous A	Lignite
Lithotype	Whole	Whole	Whole	Vitrain	Whole
Ultimate analysis (mass %)					
Carbon	83.40	69.47	66.70	69.60	65.13
Hydrogen	2.37	3.78	4.57	4.52	4.13
Oxygen	2.62	6.59	4.57	19.31	24.20
Nitrogen	0.84	1.39	1.22	0.99	0.89
Sulfur	1.03	3.75	1.90	0.65	0.57
Ash	9.74	15.02	21.04	4.93	5.08
	100.00	100.00	100.00	100.00	100.00
Proximate analysis (mass %, dry)					
Volatile matter	6.77	22.70	31.80	42.10	43.62
Fixed carbon	83.49	62.28	47.20	53.00	51.30
Ash	9.74	15.02	21.00	4.90	5.08
	100.00	100.00	100.00	100.00	100.00

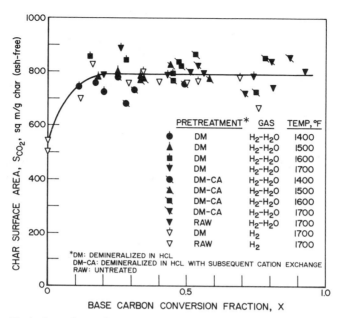

Figure 6 Variations in surface area measured in carbon dioxide (S_{CO_2}) with base-carbon conversion fraction for Montana lignite chars gasified at various conditions.

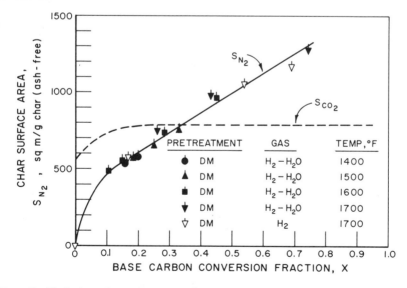

Figure 7 Variations in apparent surface area measured in nitrogen with base-carbon conversion fraction for Montana lignite chars gasified at various conditions.

249

Montana lignite chars. The characteristics shown tend to support the suggestion made previously that, at low levels of conversion, nitrogen penetration into the micropore structure is severly inhibited, but that unreasonably high apparent surface areas are obtained at higher carbon conversions because of capillary condensation.

In Figure 8 variations in surface area measured in carbon dioxide (S_{CO_2}) obtained with some other coal chars are compared with results obtained with Montana lignite chars. Although surface areas measured for char derived from anthracite, high-volatility bituminous coal, and lignite remained essentially constant during the course of conversion in hydrogen and steam-hydrogen mixtures, surface areas for the subbituminous coal char generally decreased with increasing carbon conversions during gasification in hydrogen. Interestingly, of the four coal chars tested, only the subbituminous coal char exhibited decreasing specific gasification rates during gasification with hydrogen that paralleled the decrease in surface area. This is shown in Figure 9. Although generalizations based on the kinetic behaviors exhibited by only four coal chars are not justified, these

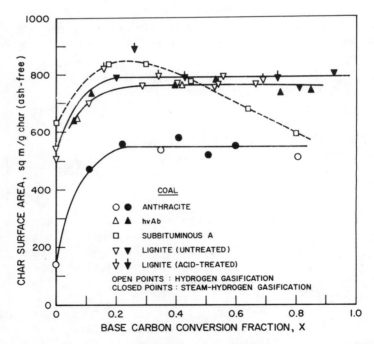

Figure 8   Variations in S_{CO_2} with base-carbon conversion fraction for different coal chars gasified in hydrogen and steam–hydrogen mixtures at $1700°F$ and 35 atms.

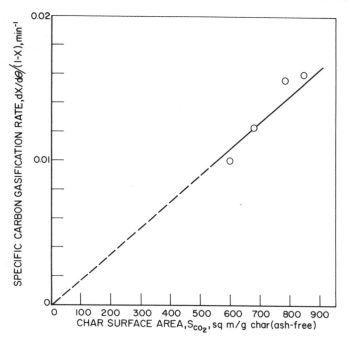

Figure 9 Relationship between specific carbon-gasification rate and S_{CO_2} for gasification of rosebud subbituminous A coal char in hydrogen at 1700°F and 35 atms.

results do suggest that a form of kinetic correlation to describe coal-char-gasification rates that is more meaningful than that derived from Equation 1 may be the following:

$$\frac{dX}{d\theta} = \lambda S k_T (1 - X) \quad [7]$$

where

λ = relative reactivity per unit of internal surface area

S = internal surface area per mass of carbon present.

With this interpretation, the constancy of the values of f_L and S_{CO_2} of anthracite, high-volatility bituminous, and lignite coal char during gasification in hydrogen corresponds to a constant value of λ for each char. For the subbituminous coal char, λ is also constant, although S_{CO_2} decreases with increasing carbon conversion, apparently because of the growth of crystallites. For the gasification temperatures used, this growth is probably unusual and not characteristic of most coal chars. This is particularly true if the first-order kinetics observed in a variety of previous studies of the gasification of a fairly large number of coal chars in hydrogen are assumed to correspond to constant values of λ and S_{CO_2} for the individual coal char tested.

Interestingly, in the kinetic model previously referred to, the value of α in Equation 1 is approximately 1.7 for a variety of gas compositions containing steam and hydrogen at elevated pressures. Empirically, this value corresponds to decreasing values of λ with increasing carbon conversions when interpreted in terms of Equation 7. This occurs even when the total internal surface areas remain constant during conversion in steam-hydrogen mixtures, as was shown in Figure 6 for Montana lignite char.

Some additional evidence was obtained in this study that can be interpreted in terms of the formulation given in Equation 7. Figure 10 shows values of S_{CO_2} obtained during gasification of a coal char derived from Brazilian metallurgical coal, at various temperatures, with hydrogen and steam-hydrogen mixtures.

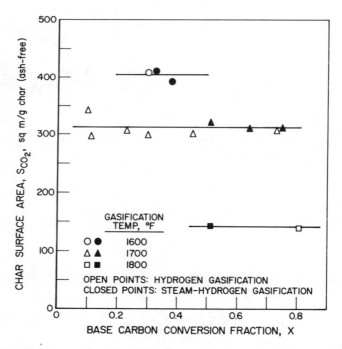

Figure 10 Effect of gasification temperature on internal surface area of carvao metallurgical coal char.

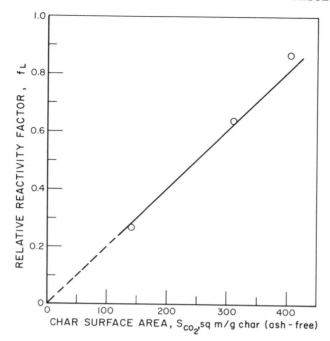

Figure 11 Relationship between reactivity factor, f_L, and S_{CO_2} for metallurgical coal char.

With this material, the char surface area, S_{CO_2}, is not a function of carbon conversion level and is the same in hydrogen and in steam–hydrogen mixtures, but decreases with increasing gasification temperature. The reactivity factor, f_L, which is characteristic of results obtained at a specific temperature, also decreases with increasing temperature and is proportional to internal char surface area, as shown in Figure 11. This particular char then can be considered to have a constant value of λ independent of temperature, conversion, or gasification medium, but it does exhibit a decreasing internal surface area with increasing temperature, a feature that probably reflects its use as a metallurgical coking coal.

Variations in Char Pore Volumes during Gasification

Figure 12 illustrates typical pore volume distributions of partially gasified coal chars. With the exception of untreated Montana lignite (curve F), the features exhibited in Figure 12 appear to be generally similar to the distributions ob-

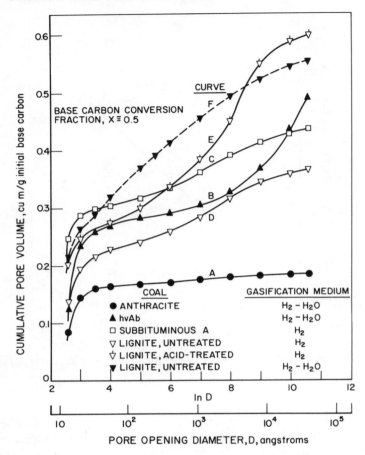

Figure 12 Typical pore volume distributions for different coal chars gasified in hydrogen and steam-hydrogen mixtures at 1700°F and 35 atms.

tained by Stacy and Walker [9] with some coal chars resulting from a fluid-bed hydrogasification. Whereas curves A-E tend to show a plateau at a pore diameter of about 55 Å, possibly indicative of the lack of development of significant transitional pores, curve F shows a significant variation in pore volume through this range of pore diameters. It may be pertinent, therefore, that the gasification rates of untreated Montana lignite char in a steam-hydrogen mixture were about 7 times faster than the largest of the gasification rates obtained with chars corresponding to curves A-E. It is thus possible that, with sufficiently large gasification rates, dynamic modifications that tend to occur within coal structures as carbon is removed are inhibited.

The plateau in pore volume variations at a pore diameter of 55 Å exhibited by most coals tested has suggested the following simplified representation of pore volume characteristics: total pore volume accessible via 'pores less than 55 Å is defined as "micropore" volume, and pore volume accessible via pore openings having diameters of 55-20,000 Å is defined as "macropore" volume. Micropore and macropore volumes obtained with different coal chars are shown in Figures 13 and 14 as a function of the base-carbon conversion fraction. Note that in these figures, volumes are represented per mass of *initial* base carbon rather than per mass of *remaining* carbon and, therefore, are proportional to volumes on a *per particle* basis. Figure 13 shows surprisingly little variation in macropore volume with increasing conversion for curves $A-E$. In viewing these results, remember that the true density of base carbon in these coal chars is about 2 g/cm^3, corresponding to a total volume of 0.5 cm^3/g of initial base carbon. Thus if the space initially occupied by gasified base carbon were added

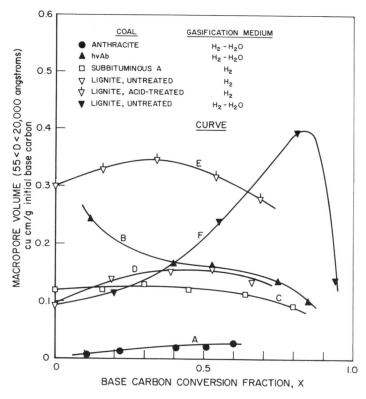

Figure 13 Relationship between macropore volume and base-carbon conversion fraction for different coal chars.

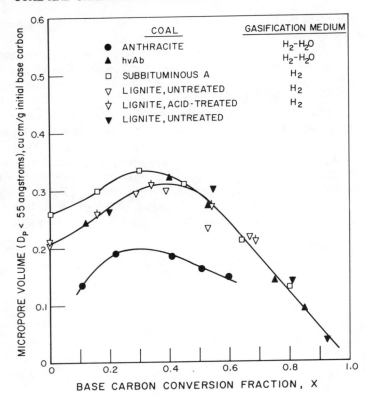

Figure 14 Relationship between micropore volume and base-carbon conversion fraction for different coal chars.

to the macropore volume, macropore volumes would increase significantly with increasing carbon conversion. The results shown in Figure 13, however, indicate that this is not generally the case, with the exception of curve F, which does show a sharp increase in macropore volume up to conversions of about 0.8.

Figure 14 shows that micropore volumes tend to initially increase with increasing carbon conversion, reach a maximum, and then decrease with increasing conversion, approaching zero at complete conversion. Interestingly, the micropore volume characteristics corresponding to untreated Montana lignite char gasified in a steam–hydrogen mixture are essentially identical to the characteristics for the other Montana lignite chars, as opposed to the behavior noted in Figures 12 and 13. Thus the rapid gasification rates that evidently affect structural transitions at a "macro" level apparently do not affect structural

RESULTS 257

transitions on a "micro" level. This is consistent with the results discussed previously, which showed an insensitivity in lignite-char surface areas to initial acid treatment or to gasification conditions.

The variations in total particle volume with base-carbon conversion measured with the mercury porosimeter are shown in Figure 15. The volumes represent the sum of solid volume plus pore volumes accessible via pore openings having diameters of less than 120 μ. As indicated in Figure 15, total particle volumes tend to decrease with increasing base-carbon conversion fraction, particularly at conversions greater than about 0.5. Because these results were somewhat unexpected when initially observed, some additional tests were conducted to obtain photographic evidence of quantitative changes that occurred

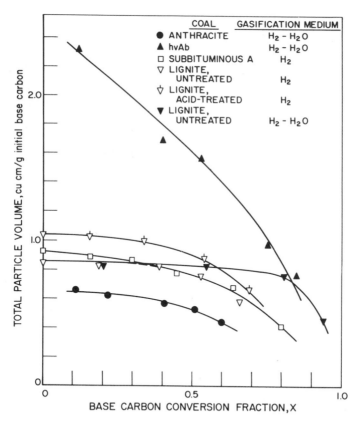

Figure 15 Relationship between total particle volume and base-carbon conversion fraction for different coal chars.

in individual external coal-char particle dimensions before and after gasification in hydrogen at 1700°F. In this series of tests, a few particles each of anthracite, high-volatility A bituminous coal, and Montana lignite were initially photographed in several orientations under optically calibrated conditions, were gasified in the thermobalance to relatively high levels of carbon conversion, and were then photographed again. Detailed examination of the photographs obtained did show a significant reduction in particle volumes, consistent with the results shown in Figure 15. The fraction of volume reduction of each type of char was independent of initial particle diameter in the approximate range 200-800 μ. This fact and the fact that external topological characteristics remained unchanged except for a diminishment in size indicated that the observed shrinkage occurred throughout individual particles and was not the result of a "shrinking-core" phenomenon.

SUMMARY AND CONCLUSIONS

The overall evidence obtained in this study suggests the tentative conclusion that gasification of coal chars with hydrogen and steam-hydrogen mixtures occurs primarily at char surfaces located within micropores. This conclusion is supported by the relationships indicated between specific gasification rates and internal char surface areas, particularly for gasification in hydrogen. With the majority of coal chars, internal surface area remains constant during gasification and is independent of gasification conditions, possibly indicating an invariance in average crystallite dimensions during the gasification process. With some coal chars, however, surface areas tend to decrease with increasing conversion or increasing gasification temperature, which would be indicative of a growth in crystallite dimensions.

Particle shrinkage occurs during coal-char gasification, due almost solely to contraction of the microporous phase (solids plus pores accessible via openings with diameters <55 Å), possibly because of the continuous reorientation of individual carbon crystallites. Macropore cavities also shrink at higher levels of conversion, but in a manner analogous to cavities in a metallic solid undergoing thermal contraction. Although accessible macropore volumes may increase somewhat during the initial stages of conversion, this increase may correspond to an increasing accessibility of the macropore cavities present initially. Although there are some significant differences in the variations in surface areas and pore volumes for various coal chars during gasification, the different coal chars tested exhibit a surprising similarity in variations in average micropore diameter with increasing base carbon conversion. This is shown in Figure 16, which plots values of the average micropore diameter, $\bar{D}$, versus the base-carbon conversion

SUMMARY AND CONCLUSIONS 259

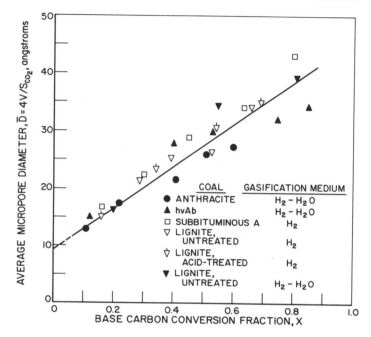

Figure 16 Relationship between average micropore diameter and base-carbon conversion fraction for different coal chars.

fraction, X. The average micropore diameter was computed from the expression

$$\bar{D} = \frac{4V}{S_{CO_2}}$$

where V is the micropore volume (accessible pore-opening diameter <55 Å).

Although the relative gasification reactivity factor, f_L, can be considered to be the product of two terms "char surface area" and "reactivity per unit of char surface area," this study has not produced enough data to evaluate correlations between these separate parameters and other coal properties. The study has shown, however, that f_L itself correlates well with the initial carbon content of raw coals for most of the coal chars tested. Lignites represent a special case, because of the catalytic effects of exchangeable cations, particularly sodium and calcium, which predominate in lignites because carboxyl functional groups are present. This catalytic effect is greater for gasification in steam–hydrogen mixtures than in hydrogen alone and tends to decrease with increasing gasification temperature.

REFERENCES

1. Dubinin, M. M., in *Chemistry and Physics of Carbon*, Vol. II., P. L. Walker, Jr., Ed., Marcel Dekker, New York, 1966, p. 51.
2. Everett, D. H., *Proc. Chem. Soc.*, 38, (1957).
3. Gan, H., Nandi, S. P., and Walker, P. L., "Porosity in American Coals," Stock No. 2414-00059, U.S. Government Printing Office, Washington, D.C., 1972.
4. Johnson, J. L., *Coal Gasification*, in *Advances in Chemistry Series*, No. 131, American Chemical Society, Washington, D.C. (1974).
5. Kaganer, M. E., *Zh. Fiz. Khim.*, 33, 2202 (1959).
6. Marsh, H., and Siemieniewska, T., *Fuel (Lond.)* 44, 355 (1965).
7. Marsh, H., and Wynne-Jones, W. F. K., *Carbon*, 1, 281 (1964).
8. Spencer, D. H. T. and Bond, R. L., *Coal Science*, in *Advances in Chemistry Series*, No. 55, American Chemical Society, Washington, D.C. 1966, p. 724.
9. Stacy, W. O., and Walker, P. L., "Structure and Properties of Various Coal Chars," Stock No. 2414-00058, U.S. Government Printing Office, Washington D.C., 1972.
10. Toda, Y., "Study on Pore Structure of Coals and Fluid Carbonized Products," report No. 5 of the National Institute for Pollution and Resources, Japan, 1973.
11. Toda, Y., Hatauci, M., Toyoda, S., Yoshida, Y. and Honda, H., *Fuel*, 50 (2), 187 (1971).
12. Walker, P. L., and Patel, R. L., *Fuel*, 49 (1), 91 (1970).

5 The Use of Catalysts in Coal Gasification

INTRODUCTION

Existing and projected shortages of natural gas in the United States have stimulated extensive research to develop the technology for commercial production of synthetic high-Btu gases of pipeline quality. In this general effort, coal gasification has been given particular emphasis because of the extensive availability of coal as a natural resource in this country, and several large-scale process-development programs are currently being conducted under government and industrial support. The major thrust of these programs has been to utilize modern technologies to develop processes with significantly improved efficiencies as compared with classical European gasification systems. Data available from the different programs currently at various stages of development already indicate that the objectives of improved efficiency and performance will indeed be achieved in the relatively near future. This evidence also indicates, however, that practical limitations still exist to efficiency levels obtainable solely through manipulation of process configurations and the use of modern gas-solid contacting systems, and that these limitations are due primarily to the chemistry inherent to coal gasification in the absence of major catalytic influences. The fact that these limitations are not theoretical or thermodynamic in nature but result from relative and absolute chemical kinetic characteristics has created considerable interest in the use of catalysts in gasification systems as a basis for further significant improvement in process efficiencies.

A consideration of catalytic effects in coal-gasification systems is unavoidable, even if catalysts are not added to the coal, because mineral and trace inorganic constituents present in the coals can have catalytic influences on gasification reactions. In addition, the gasification systems being developed require purification of the gasifier product gases including catalytic water-gas shift to adjust carbon monoxide and hydrogen concentrations for final catalytic

---

Adapted from a paper presented at the 170th Meeting of the ACS/DFC, Chicago, August 24-29, 1975, Fuel Chemistry Division, ACS, Preprints 20(4): 31 (1975), with permission.

methanation to produce a pipeline-quality gas. Although major improvements in process performances can be realized by improved purification catalysts and through increased understanding of the coal's mineral constituents, there is ample evidence that even greater improvement in process performance could result from adding catalysts to the coal mass. Because of this potential, it is useful to review certain aspects of current technology relevant to coal gasification catalysis, to aid in assessing future research needs.

GENERAL CHARACTERISTICS OF COAL-GASIFICATION SYSTEMS

Figure 1 represents the overall flow sheets for many coal-gasification process concepts directed toward methane production. Steam, oxygen, and coal, fed to a gasifier system, are converted to a synthesis-gas mixture containing substantial methane at elevated pressure, and to residual carbon monoxide and hydrogen, which must be converted to methane catalytically to produce a pipeline-quality gas. The main practical objectives in the engineering of such systems are to reduce coal and oxygen feeds, reduce reactor sizes, minimize purification operations, and simplify the physical complexity of the gasifier elements. The prime obstacle in attaining these objectives is the relatively high gasification temperature required to promote the desired carbon conversions within reasonable reactor size limits. Unfortunately, these required temperature levels inhibit direct methane formation in the gasifier and, instead, promote carbon monoxide and hydrogen formation, which favors increased oxygen, feed-coal, and purification requirements, thereby decreasing overall process thermal efficiencies.

A more detailed representation of reaction processes occurring within a gasifier is given in Figure 2. The total gasification process can be broadly

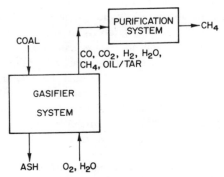

Figure 1 Coal-gasification flow sheet.

GENERAL CHARACTERISTICS OF COAL-GASIFICATION SYSTEMS 263

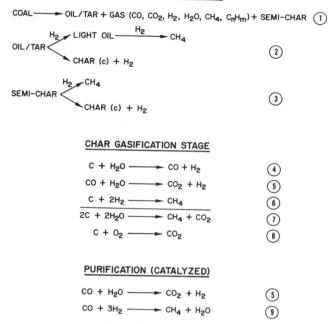

Figure 2 *Overall gasification reaction steps.*

separated into two main reaction stages: "coal gasification" and "char gasification." The coal-gasification stage corresponds to reactions related to devolatilization that occur during initial heat up of a coal. These reactions occur quite rapidly, in seconds or less at elevated temperatures. The char-gasification stage corresponds to the much slower gasification of the coal char after devolatilization is essentially completed.

Coal-stage Reactions

During the coal-gasification stage, up to about 50% of the original coal mass can be gasified. Some of the overall transitions that contribute to conversion and product yield in this reaction stage can be classified as follows. Reaction ① (in Figure 2) corresponds to pyrolysis processes occurring below 700°C, in which oils and tars evolve, most of the coal oxygen is liberated as carbon oxides and steam, and relatively small amounts of gaseous hydrocarbons are formed. There is little evidence to indicate that gas composition has a major effect on this reaction, although higher hydrogen partial pressures tend to increase steam yields somewhat, relative to carbon oxide yields. Higher heat-up rates and dilution of coal concentrations lead to higher oil-tar yields, but for most current processes the coal heat-up rate is not a particularly flexible parameter, and initial oil-tar yields are largely a function of coal type. The oils and tars that initially evolve from the coal mass can react further, depending on subsequent conditions in the gasifier. This additional activity, symbolized by Reaction ②, can result either in the formation of char, through cracking, or of lighter oils, and possibly even methane, through further hydrogenation. This latter process is favored by increased hydrogen partial pressure, increased temperature, and dilute coal concentrations.

The transitions indicated by Reaction ③ are particularly important in processes directed toward methane production. It has been shown that when coals are devolatilized in the presence of hydrogen they exhibit an exceptionally high, though transient, reactivity for methane formation. Although this transient period lasts only for seconds or less at temperatures above about 800°C, a substantial fraction of the coal's carbon can be gasified to methane at elevated hydrogen partial pressures. In the Institute of Gas Technology's (IGT) HYGAS® Process, for example, approximately half the total methane formed in the gasifier results from this reaction step at a total operating pressure of 1000 psi. Although the mechanism for this reaction is not well known, recent work at the IGT [15] has indicated that the reactive solid state is strongly related to the secondary pyrolysis associated with hydrogen evolution during the final stages of char formation. This reaction initiates at about 600–700°C and is essentially consecutive to the main pyrolysis reactions indicated by Reaction ①.

Char-stage Reactions

After the coal-gasification stage is completed, the remaining char is much less reactive than the original coal and requires extended residence time for the further gasification reactions indicated in the char-gasification stage. The primary process objective of char gasification is Reaction ⑦, which is almost thermally neutral. Unfortunately, this overall reaction is not kinetically feasible in uncatalyzed systems. The actual reaction path corresponds more closely to a three-part synthesis: hydrogen and carbon monoxide formation by Reaction ④, shift by Reaction ⑤, and methane formation by Reaction ⑥. Reaction ④ is endothermic and does not normally occur at reasonable rates below about 800°C with bituminous coals. For most processes, in fact, temperatures of about 1000°C or greater are necessary to obtain the required conversions with reasonable reactor sizes. This level of temperature, however, does not favor the exothermic Reaction ⑥, which is relatively slower than Reaction ④ at elevated temperatures. Some processes improve this situation somewhat by the use of multizoned reactor systems in which hydrogen generated in a high-temperature zone by Reaction ④ subsequently reacts with carbon in a lower temperature zone to form methane by Reaction ⑥. Even with this procedure, the overall

GENERAL CHARACTERISTICS OF COAL-GASIFICATION SYSTEMS 265

gasification process is still endothermic. Additional heat to sustain the reactions—usually achieved by further combustion of the char with oxygen—requires increased feed coal and oxygen and increased catalytic methanation to produce a given amount of final methane product, all of which decrease the process thermal efficiencies.

The potential for approaching the stoichiometry indicated by Reaction ⑦ is possible only as a result of the action of catalysts that preferentially enhance either direct formation of methane, such as in Reaction ⑥, or indirect formation of methane through a reaction such as Reaction ⑨ occurring in the gasifier itself. Thermodynamic considerations, however, dictate that even with optimal catalytic action, substantial improvements in process efficiencies can occur only through operation of the gasifier at temperatures much lower than those normally required in the absence of catalysts.

An illustration of the effects of temperature on gasification process performance is shown in Figure 3 for an idealized gasification system. The characteristics shown in Figure 3 were calculated for adiabatic operation of a backmixed

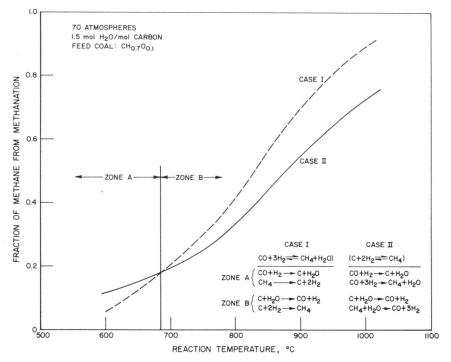

Figure 3 Effect of temperature on methanation requirements.

gasifier at 1000 psi, with a feed steam: coal ratio of 1.5 mol/mol, for complete gasification of a coal having the composition $CH_{0.7}O_{0.6}$. Feed and product-stream temperatures were assumed to be 25°C, feed oxygen requirements were computed to satisfy adiabatic operation, and the water-gas shift reaction [Reaction ⑤] was assumed to be in equilibrium at the gasifier temperature. With these assumptions, two cases were considered: case I assumes that Reaction ⑨ is at equilibrium in the gasifier, corresponding to a catalytic effect on secondary methane formation reactions; case II assumes that Reaction ⑥ is at equilibrium, corresponding to a direct catalytic effect on gasification. Both cases show a substantial reduction in required external methanation with decreasing gasifier temperature. As indicated previously, decreasing external methanation corresponds directly to decreasing feed oxygen and coal requirements per mole of total methane in product gases resulting from external methanation. Above 685°C (zone B), the assumption of equilibrium for Reaction ⑥ (case II) results in somewhat improved efficiencies as compared with case I, which assumes equilibrium for Reaction ⑨. At about 685°C, however, the total system is at equilibrium, including the steam-carbon reaction [Reaction ④]; therefore, this temperature represents an optimum condition for the idealized system considered. Although the specific temperature value is dependent on the specific values of the other operating parameters assumed, this general level of temperature is characteristic of the conditions required to best approach the stoichiometry of Reaction ⑦ through the use of catalysts.

EXPERIMENTAL STUDIES RELATED TO USE OF CATALYSTS IN COAL GASIFICATION

Considerable experimental work has been conducted to study the effects of catalysts on the gasification of carbonaceous solids. Although many of these studies have used relatively pure carbons or have been conducted at conditions not directly related to coal-gasification systems, much of the empirical information obtained is of value in assessing future research needs in developing the technology necessary for commercialization of catalyzed coal-gasification systems.

Experimental Studies Related to the "Coal-gasification Stage"

Most work related to reactions in the coal-gasification stage has been conducted with the objective of catalytic enhancement of hydrocarbon liquid and gas production through hydrogenation at elevated pressures. These pressure levels have usually been far in excess of pressures normally considered in coal-to-methane gasification processes, but the results are of interest because of the possible economic advantages of combined hydrocarbon liquid and gas production from coal, as well as the possibility of hydrocarbon gas formation through further hydrogenation of the liquids according to Reaction ②.

Experimental studies have been conducted using batch autoclave systems; fixed-bed, flowing gas systems; and continuous, dilute-solids-phase transport systems, at hydrogen pressures of 500-6000 psi, and at temperatures of 250-1000°C. Catalysts have been added to the coal mass by physical mixing or have been impregnated within the individual coal particles by evaporation from solution. Although the effects of a large variety of catalysts have been evaluated, the major emphasis during recent years has focused on the use of metal halides and molybdenum catalysts.

Several programs have been conducted by the U.S. Bureau of Mines since the early 1960s to study coal hydrogenation catalyzed by ammonium molybdate in a fixed-bed, flowing gas system [7, 12, 13]. A variety of coals were used, with impregnated catalyst concentrations of up to 1%, which appears to be an optimum concentration. In the experimental system usually employed, a fixed bed of coal solids was placed in a small-diameter tubular reactor that was heated electrically by passing current through the tube walls, achieving a heat-up rate of about 5°C per second above 300°C. Experimental results were characterized by "time at temperature," flowing gas residence time through the fixed bed, maximum temperature, and pressure.

Substantial hydrocarbon liquid yields were obtained with the ammonium molybdate catalyst, ranging as high as 60% of the feed coal at 6000 psi. Increased hydrogen pressure favored increased liquid yields, and under most conditions studied, maximum liquid yields were obtained at 500-600°C, although the liquids formed at this temperature level were usually heavy in nature. At higher maximum temperatures, liquids tended to become "lighter" in nature, approaching BTX quality at about 700-800°C. It was noted, however, that (particularly at higher temperatures) the flowing-gas residence time is a critical parameter in determining both liquid yield and liquid quality. In one series of experiments at 800°C and 6000 psi with a New Mexico subbituminous coal [7], it was observed that for "zero" coal residence time (heat up to reaction temperature, then quench) liquid hydrocarbon yields increased from about 10% to about 40% of the feed coal for a corresponding decrease in gas residence time from 25 sec to 5 sec. For this range of residence times, the hydrocarbon gas yield was not substantially affected, and the oil formed was light in color and contained a large quantity of distillable oil. For a further decrease in residence time to 2 sec, however, although the oil yield was about the same as that obtained at 5 sec, the oil was dark, viscous, and contained an appreciable quantity of asphaltenes. These results suggest that the primary oils evolving from coals (generally below 600°C) are heavy in nature, but subsequent exposure of the gasified material to elevated temperatures can lead to light aromatic oil formation in the presence of gaseous hydrogen. Too long an exposure, however, particularly in a fixed coal bed, can lead to resolidification through cracking reactions, a process promoted by the coal solids themselves. In a dilute-solids phase, with a short residence time and transport reactor system, however, significantly different behavior occurs. Schroeder, Stevenson, et al. [26] have reported that, for the hydrogenation of an ammonium molybdate catalyzed coal in such a system, the decrease in liquid hydrocarbon yield associated with extended exposure time does not result in a decrease in total coal conversion, suggesting that liquids and tars convert to hydrocarbon gases rather than resolidifying. This is an important point, because when catalysts are used in an integrated continuous coal-gasification process, it would be quite reasonable to conduct the initial reaction stages in a dilute-solids-phase transport reactor.

Reported experimental studies using ammonium molybdate contain little or no information on the important aspects of catalyst recovery and possible side reactions, other than one hypothesis that some molybdenum may be partially removed from a flowing system as volatile molybdic oxide [7]. Because of the relatively high cost of ammonium molybdate as a raw material, its loss during hydrogenation should be carefully evaluated in future studies. Although ammonium molybdate appears to be somewhat more active as a hydrogenation catalyst than many other materials, there has been considerable interest in the use of metal halides because of their generally increased availability and in some cases their greater ease of impregnation and dispersion in coal particles. The merits of metal halides as coal hydrogenation catalysts were shown during the 1950s by Weller, Pelipetz, et al. [32, 33], who studied the relative effects of a variety of catalysts on hydrocarbon liquid production during hydrogenation of a high-volatility bituminous coal at 450°C and 1000-3700 psi. This study was conducted with a batch autoclave system and residence times at temperature of about 1 hr, using mixed and impregnated catalysts. It was found that impregnated metal halides ($ZnCl_2$, $SnCl_2$) at up to 1% concentrations appeared to be most effective in the formation of hydrocarbon liquids, and a simplified model was suggested to explain experimental results. This model, illustrated in Figure 4, indicates that liquid hydrocarbon formation results from initial fragmentation of the coal structure, followed by competitive reactions in which fragments either poly-

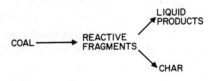

Figure 4 Model for liquid hydrocarbon formation.

merize to char or become stabilized by hydrogen to form liquid products. With this model, it was postulated that catalysts can influence both the initial coal fragmentation step and the hydrogen stabilization step. It is noteworthy that this simplified qualitative model has not been substantially supplemented with greater mechanistic detail during the last 25 years, although recently Matsura, Bodily, et al. [23] conducted studies which indicated that metal halides form surface compounds with coals. These investigators also report that the tendency to form surface compounds is, in fact, reflected in the Lewis acidity of the materials, although the fact that a surface compound is formed is not a sufficient criterion for catalytic reactivity.

Studies have also been conducted at the University of Utah to investigate the effects of a variety of impregnated catalysts on the hydrogenation of several coals in a dilute-solids-phase transport reactor [11, 36]. These studies—of particular relevance to continuous coal-gasification systems—also showed that metal halides, including $ZnCl_2$ and $SnCl_2$, are very effective as promoters of liquid hydrocarbon formation. Of the metal halides tested at temperatures up to 700°C and pressures to 2500 psi, a preference was indicated for $ZnCl_2$ at an optimum concentration of 6%, because of greater economic availability, ease of impregnation, and greater potential for ultimate recovery in a gasification system. In one of the studies, high conversions were obtained for coals impregnated with an optimum $ZnCl_2$ concentration of about 6%, for solids residence times less than 10 sec and gas residence times less than 0.03 sec [36]. In this study pressures ranged 1500-2500 psi and temperatures, 625-700°C. Within this range of conditions, liquid yields increased only 625-650°C (20-40%) and remained constant up to 700°C, whereas gaseous yields were relatively constant over the total range (ca. 15%). Only slight variations in both liquid and gaseous yields with variations in pressure were obtained for the range studied. Even with $ZnCl_2$, ultimate recovery after gasification was noted to be a practical problem because of its conversion to less extractable forms such as zinc and ZnS.

Although many catalysts have been found to enhance liquid hydrocarbon formation during reaction in the coal-gasification stage, very little enhancement of direct gaseous hydrocarbon formation has been observed in the studies conducted. One interesting exception to this trend was obtained in a study in 1968 by Kawa, Friedman, et al. [18] using a batch-autoclave system to investigate liquid and gaseous yields from different coals for hydrogenation at 250-450°C and 4000 psi, using $AlCl_3$ as a catalyst. The $AlCl_3$, which was physically mixed with the coal charge, is more volatile than $ZnCl_2$ or $SnCl_2$ and could be completely vaporized at the experimental test conditions employed. For reaction times of 1 hr with a 50:50 coal-$AlCl_3$ mixture, the gaseous hydrocarbon yields from a high-volatility A bituminous coal at 450°C were approximately 68% of the raw coal, with negligible benzene-soluble oil formed. Significantly, 96% of the hydrocarbon gas was comprised of methane and ethane. Even at 300°C, a hydrocarbon gas yield of about 40% was obtained, although at this condition about 15% yield of oil resulted. A major problem with the $AlCl_3$, however, appears to be from side reactions in which the $AlCl_3$ is converted to Al_2O_3 by

$$O(coal) + H_2(gas) \longrightarrow H_2O(gas) \qquad [1]$$

$$(AlCl_3)_2(gas) + 3H_2O(gas) \longrightarrow Al_2O_3 + 6HCl(gas) \qquad [2]$$

This effect was noted to be most severe with coals of higher oxygen content and was reflected in decreased hydrocarbon yields because of the decreased net $AlCl_3$ available. This is illustrated in Figure 5, which shows hydrocarbon yields versus coal oxygen content obtained for 50:50 coal–$AlCl_3$ charges for tests conducted with other coals and carbonaceous solids at 450°C. With the exception of anthracite, the hydrocarbon yield is apparently inversely proportional to the coal's oxygen content. Point A in Figure 5 corresponds to the

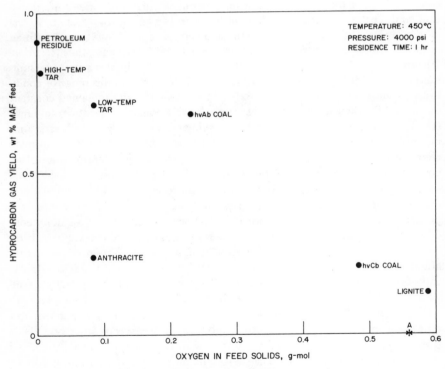

Figure 5   Effect of coal oxygen on gas yield.

amount of oxygen required for complete stoichiometric conversion of the $AlCl_3$ charged to form Al_2O_3. The results obtained with $AlCl_3$ suggest a direct catalytic attack on the initial coal strucutre in a manner not symbolized by the coal-gasification-stage reactions presented previously, and an explanation of its effect does not appear reasonable in the general terms used to explain catalytic effects with other metal halides. The potential for such substantial gaseous hydrocarbon formation at such low temperatures is, however, an extreme inducement for a further search for catalysts or catalyst systems to achieve such a dramatic conversion, without undesirable side reactions.

The main body of experimental evidence pertaining directly to those catalytic reactions in the coal-gasification stage that promote the production of hydrocarbon liquids and light hydrocarbon gases apply primarily to systems using pure hydrogen feed gas and to hydrogen pressures significantly higher than would normally be used in an integrated process where hydrogen is produced by steam–oxygen gasification of the residual char. Therefore, one potential area for future research is the evaluation of catalytic effects on reactions occurring in the coal-gasification stage, using synthesis-gas mixtures representative of product gases from char gasifiers, at total pressures of 1000 psi or less. In general, it would appear that such studies would best be conducted with continuous dilute-phase transport systems using impregnated catalysts or catalyst combinations, and that the problems of side reactions and eventual catalyst recovery should be strongly evaluated.

Experimental Studies Related to the Char-gasification Stage, or to Both Coal- and Char-gasification Stages

Since the 1920s a variety of studies have investigated the effects of catalysts on the gasification of carbonaceous solids, with the objective of improving processes for the production of water gas, producer gas, or hydrogen as sources for ammonia. More recently, studies have also been conducted in connection with gas-cooled nuclear reactors [2, 3, 5, 6, 19-22, 25, 27-31]. Although most work with these objectives has been conducted at low pressures with cokes, charcoals, and graphite, much of the information obtained with the large number of catalysts evaluated is of at least qualitative value in assessing their potential effects on char gasification for methane production. Catalysts surveyed in earlier work include metals, metal oxides, metal halides, alkali carbonates, and iron carbonyls, with particular emphasis given to K_2CO_3, Na_2CO_3, KCl, NaCl, CaO, and transition metals. Some of the general conclusions of the work are as follows:

1. Relative catalytic effects decrease with increasing gasification temperature.
2. In the gasification process, catalysts are generally more effective with gases containing steam than with hydrogen alone.

3. There is usually an intermediate optimum catalyst concentration beyond which either negligible or negative effects result.
4. The relative effects of different catalysts can differ at reaction conditions.
5. The specific methods and conditions employed for catalyst impregnation can significantly affect subsequent gasification reactivity.
6. Catalyst impregnation is more effective than physical mixing with the carbon.

With respect to this last point, the method of catalyst addition is very important to coal-gasification processing. The advantages of catalyst impregnation must be economically weighed against the corresponding requirements for catalyst-addition operations and catalyst-recovery problems, which are likely to be more severe than with physically mixed catalyst-coal systems. The use of physical mixtures of catalysts and coal in laboratory studies presents a particular problem when the results are interpreted in terms of projected continuous gasifiers, because the transfer mechanism required for the catalyst to interact in the gasification step may depend on the physical nature of catalyst-coal contacting. Such contacting would be quite different for the fixed beds frequently used in laboratory studies than for the fluidized beds more likely to be used in large-scale systems.

Available information most directly relevant to catalysis of coal-char gasification has been obtained only within the last several years. In one such investigation, conducted at Case Western Reserve University [9, 34], a thermobalance apparatus was used to study the effects on coal-char gasification kinetics of approximately 20 impregnated catalysts. In contrast to integral gas-conversion systems employed in the vast majority of catalyst studies, the thermobalance can be used to study coal-char gasification kinetics under constant environmental conditions, a situation which greatly facilitates detailed quantitative evaluation of catalytic effects. In thermobalance studies reported by Gardner, Samuels, et al. [9] and by Wilks, Samuels, et al. [34], catalytic effects were investigated at 850-950°C and 500-1000 psi for gasification in either hydrogen or steam. Most tests used a 5% catalyst concentration. It was concluded that although a few catalysts are as effective as the alkali metals in promoting gasification rates, none was found to be more active. With 5% $KHCO_3$, gasification rates in hydrogen at 500 psi and 950°C are increased twofold, whereas in steam, gasification rates are increased about threefold at the same parameters. More recently, data have been obtained indicating that catalyst combinations, such as alkaline-earth metals with cobalt molybdate, are even more effective than single catalysts in enhancing gasification rates [8], suggesting that a study of catalyst combinations would be a promising area for future research.

A thermobalance apparatus is being used at IGT to study the catalytic effects of exchangeable cations, such as sodium or calcium, on lignite-gasification reactivities [16]. This study was prompted by the fact that, although lignite chars are inherently more reactive than bituminous chars for gasification, when treated in acid (e.g., HCl), their gasification reactivity is substantially reduced and quite comparable to the reactivities of bituminous chars. It has been postulated that the inherent high reactivities of lignites is due to the catalytic influences of sodium or calcium combined with carboxyl functional groups in the organic structure of lignites. When treated in acid, however, the cation can be removed by the following exchange reaction:

$$-COONa + H^+ \rightleftharpoons -COOH + Na^+ \qquad [3]$$

Since the concentration of carboxyl functional groups decreases significantly with increasing coal rank, this inherent catalytic effect is not significant in bituminous coal chars.

In studies conducted at IGT, the reversibility of Reaction was used as a basis to systematically investigate the effects of exchangeable cation concentration on lignite char reactivities. Lignites were initially treated in hydrochloric acid to remove inherent calcium or sodium; controlled amounts of calcium or sodium were then added to the lignite by subsequent exchange through reverse of Reaction ③. Experimental results showed that for a 3% sodium concentration, a twentyfold increase in gasification rates was obtained at 760°C and 500 psi in a steam-hydrogen mixture, as compared with rates obtained with the lignite when containing no exchangeable cations (acid treated only). In general, gasification rates increased with increasing catalyst concentration, at least up to 3%, and similar results were obtained with both calcium and sodium at comparable mass concentrations, for gasification in steam-hydrogen mixtures. Relative catalytic effects decreased with increasing gasification temperature, such that with steam-hydrogen mixtures at 930°C, only a sevenfold increase in gasification rates was obtained with 3% catalyst addition. For gasification in pure hydrogen at 930°C, the catalytic effects were significantly less than with steam-hydrogen mixtures. In addition, sodium was found to be about 4 times more effective than calcium for gasification in pure hydrogen.

Attempts to add sodium and calcium to bituminous coals have been unsuccessful, presumably due to the lack of carboxyl function groups. It is not known whether the exceptionally large catalytic effects obtained with the organically combined sodium and calcium in lignites are due to the particular chemical state of the catalytic complexes or to the greater dispersion that is likely compared with simple deposition on internal surfaces by solution evaporation, but future studies to determine methods to promote organic complexes between coal and catalyst would appear to be promising.

Some recent studies have been conducted at several laboratories to investigate the gasification yields that can be obtained with physical catalyst-coal mixtures under conditions of integral gas conversion, at elevated pressures. In one such study conducted by the U.S. Bureau of Mines [4], the effects of physically mixed catalysts were investigated in a continuous, 4-in.-diameter, fluidized-bed reactor used in the development of the "synthane process." Additives included 2%, 5%, and 10% dolomitic limestone, 2% and 5% hydrated lime, and 5% lignite ash. For operation at 600 psi with a steam–oxygen feed, at 875–950°C, total carbon conversions increased by an average of about 10% with the added materials. This increase in conversion is equivalent to about a 50°C decrease in gasifier temperatures, compared with temperatures required for a given coal throughput in the absence of the additives. It was also noted that, with the majority of additives tested, no slag formation occurred, suggesting that the gasifier could be operated at higher temperatures, if desired, to achieve higher throughputs.

In another Bureau of Mines study, reported by Haynes, Gasior, et al. [10], the effects of approximately 40 different catalysts on yields obtained by the gasification of a high-volatility bituminous coal with steam at 850°C and 300 psig were evaluated. In this study, 5% catalyst concentrations were used in physical mixtures with the coal charge. Experimentally, a fixed-bed, flowing gas system was employed, and total gaseous yields were reported for 4-hr test periods. It should be noted that, as raw coals were used in this study as well the Bureau of Mines study discussed earlier [4], total gas yields result from reactions potentially occurring in both coal- and char-gasification stages.

Haynes found that, as in earlier studies, the best overall reactivities were obtained with alkali catalysts, particularly K_2CO_3 and KCl. With these materials, total carbon conversions of about 80% were obtained during the 4-hr test period, compared with a reference carbon conversion of less than 50% obtained in the absence of any catalyst. For the most part, increased reactivity led to increased formation of carbon oxides and hydrogen, relative to increases in methane formation. An exception to this trend was obtained with an unactivated Raney nickel catalyst sprayed onto the surface of a metal tubular insert in the coal bed. In this case, total methane yields increased by about 24%, although total carbon conversion only increased by about 10%, indicating that the major influence of the catalyst was in methanation of the carbon monoxide and hydrogen formed from direct gasification. The sprayed Raney nickel catalyst, however, decreased in reactivity with continued use, and loss of reactivity was estimated at 20 hr of operation. This loss of reactivity under the relatively severe conditions existing in the gasifier was attributed both to "flaking off" of the catalyst and to sulfur poisoning, which is currently one of the major problems in the use of nickel-based catalysts in coal gasifiers.

In evaluating the significance of catalysts in coal-gasification systems with gases containing both steam and hydrogen, it is important to distinguish between two major aspects: (1) the increase in total carbon-gasification rate and (2) the relative increase in hydrocarbon production. This latter aspect is particularly important in reflecting the increased thermal efficiencies in a gasification process and in many instances is probably an overriding factor in determining the practicality of using catalysts in coal gasifiers. For the Bureau of Mines studies discussed earlier [4, 10], it is thus significant that, although additives generally tended to increase the total carbon conversion rates, with one exception, the increased carbon conversion apparently corresponded to a decrease in thermal efficiency for the overall gasification step. This can be shown quantitatively if it is assumed that incremental variations in total carbon conversion and methane yields due to catalytic effects correspond to variations in the extent of conversions of the following reactions:

$$C + 2H_2O \longrightarrow CO_2 + 2H_2 \qquad [4]$$

$$C + 2H_2 \longrightarrow CH_4 \qquad [5]$$

With this assumption, the experimental yields can be used to compute incremental heats of reaction relative to the uncatalyzed system, to determine whether there is an incremental heat excess or deficiency resulting from the use of catalysts corresponding to an increase or decrease in thermal efficiency. In Figure 6 the values of relative methane yield versus total carbon conversion are plotted for some of the more extreme results reported by Haynes and co-workers. The dashed line represents the condition for no variation in thermal efficiency with variations in total carbon conversion, relative to operation with no catalyst. It is seen that, except for the sprayed Raney nickel catalyst, operation with all other catalysts results in decreased thermal efficiency; this decrease becomes more severe with greater the activity of the catalyst. This effect is generally consistent with results obtained in other investigations where catalysts observed to promote the greatest increase in gasification rates also tended to preferentially promote carbon oxide formation relative to methane formation. These results suggest the possible advantages of use of the more reactive catalysts at lower temperatures to enhance methane yields rather than at higher temperatures to reduce reactor sizes.

The use of catalysts at lower temperatures has been the subject of several studies conducted at the University of Wyoming [1, 14, 35] having the objective of promoting the overall optimal reaction

$$C + 2H_2O \longrightarrow CH_4 + CO_2 \qquad [6]$$

In one series of experiments, the gasification with steam of low-rank coals mixed with K_2CO_3 and commercial nickel catalyst was studied in a fixed-bed, flowing gas system at 650°C. It was found that, with this system, it was indeed possible, even a low pressure (30 psi), to produce a gas consisting primarily of CO_2 and CH_4, with optimum catalyst concentrations of about 1 g of nickel catalyst/1 g

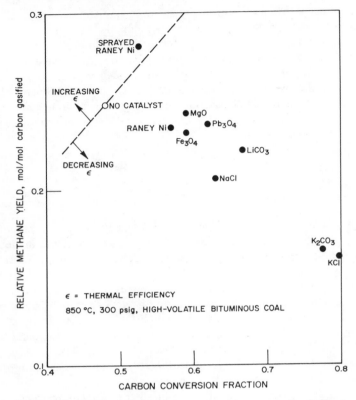

Figure 6   Thermal efficiency variations due to catalyst effects.

of coal and 0.2–0.25 g of $K_2CO_3/1$ g of coal. On a CO_2-free basis, product gases were consistently obtained with heating values of about 850 Btu/scf. At 650°C, however, these yields were obtained for total test periods of about 7.5 hr and for total carbon conversions of less than 50%. Such results suggest that, for operation at this low temperature, a much more active catalyst system or coal–catalyst contacting technique would be required for acceptable large-scale operation.

Two other major problems also inherent to this system are catalyst recovery and the loss of catalyst activity. For optimum reaction conditions, only about 80% of the potassium charged was recovered by a 1-hr water wash at room temperature, a relatively serious condition in view of the large amounts of K_2CO_3 charged. In addition, relatively large fractions of the coal's sulfur combined with the nickel catalyst during gasification. Although this did not result in noticeable decreases in reactivity during batch charge tests in view of the large amounts of nickel catalysts employed, extrapolation of these effects indicated that replacement of completely sulfided catalyst would be economically unfeasible and that some means of catalyst regeneration would have to be employed in commercial application. Several approaches to dealing with catalyst poisoning by sulfur have thus far not been completely satisfactory, and it has been suggested that perhaps a combination of the use of sulfur scavengers, catalyst regeneration, more powerful sulfur-resistant catalysts, and prior sulfur removal may be required as a final solution.

SUMMARY AND CONCLUSIONS

A large number of experimental studies, such as those discussed, have shown that the use of catalysts in coal-gasification systems has many advantages. In individual studies, such desirable results have been demonstrated as increased gasification rates, improved product yields, and decreased sintering and agglomeration characteristics. It is also evident, however, that a number of problems exist in the use of catalysts in coal gasifiers (particularly in systems that produce the most extreme desirable effects) that must be resolved before optimum potentials can be practically realized. Some of the more obvious problems relate to catalyst recovery, undesirable side reactions and catalyst poisoning, and the possible undesirable effects of vaporized catalysts or catalyst derivatives on gasification-purification systems.

One of the particularly significant trends indicated from available studies is that the catalysts that most significantly enhance *total* gasification rates in steam-hydrogen systems preferentially lead to the formation of carbon oxides relative to methane. This behavior has suggested the use of catalyst combinations—with one type of catalyst to promote total gasification rates at lower temperatures, and another type to promote the secondary reactions leading to methane formation that are thermodynamically favorable at lower temperatures. An example of one study using this approach has been discussed, where K_2CO_3 and Ni catalysts were physically mixed with the coal charge. Unfortunately, in this study, although the kinetics of secondary reactions to form methane were reasonable, the total carbon gasification rates were unacceptably low for commercial application. Although this result would probably be improved significantly by impregnation of the K_2CO_3 catalyst, which enhances total carbon-gasification rates, it is also probable that a more active catalyst or catalyst combination will be necessary to sufficiently increase total gasification rates at the low temperature necessary for optimum relative methane yields.

In general, the observations discussed in this paper are consistent with other recent evaluations [17, 24] that conclude that there is a substantial need for additional research in general areas related to the use of catalysts in coal gasification. Specifically, these needs include the development of new catalysts and catalyst systems, with particular emphasis on detailed investigations of mechanisms involved in the interactions of catalysts with coal- or char-gasification processes. Such detailed considerations would ideally involve a much more fundamental characterization of reaction systems than has been done in the bulk of previous studies. For example, particular attention should be directed toward characterization of the coals and coal chars used, not only in terms of elemental composition or rank type, but also of petrographic composition and internal physical properties such as surface area and pore volume distribution. The dispersion characteristics, chemical state, and physical properties of the catalysts employed should also be determined, not only in the raw coal, but also during the course of gasification.

With apparently desirable catalyst systems, detailed kinetic studies would be needed to provide the information necessary to fully evaluate their application in an integrated commercial system over the range of environmental conditions to which the catalyst would typically be exposed. Such kinetic studies should include not only major reactions, but also side reaction and reactions pertinent to catalyst recovery. Although these objectives are overwhelming in some respects, such research could progress at a much more rapid pace than previous developments through sufficiently large, integrated efforts and support, utilizing modern sophisticated analytical techniques. In view of the obvious potential improvements that efficient use of catalysts may bring about in the performance of coal-to-methane conversion processes and the urgent need to develop these processes, such an intensified, broad program to study coal-gasification catalysis appears to be quite justified.

REFERENCES

1. Cox. J. L., and Sealock, L. J., Jr., "Sulfur Problems in the Direct Catalytic Production of Methane from Coal–Steam Reactions," in *Proceedings of the 167th National Meeting of the American Chemical Society, Division of Fuel Chemistry*, Vol. 19, No. 1, 1974, pp. 64–78.
2. Dent, F. J., Blackburn, W. H., and Millett, H. C., "The 41st Report of the Joint Research Committee: The Investigation of the Use of Oxygen and High Pressure in Gasification–Synthesis of Gaseous Hydrocarbons at High Pressure. Part III. Detailed Account of Investigations," *Transact. Inst. Gas Eng.*, 87, 231–287 (1937).
3. Dent, F. J., Blackburn, W. H., and Millett, H. C., "The 43rd Report of the Joint Research Committee: The Investigation of the Use of Oxygen and High Pressure in Gasification–Synthesis of Gaseous Hydrocarbons at High Pressure. Part III. Detailed Account of Investigations," *Transact. Inst. Gas Eng.*, 88, 150–217 (1938).

4. Forney, A. J., Haynes, W. P., Gasior, S. J. and Kenny, R. F., "Effect of Additives Upon the Gasification of Coal in the Synthane Gasifier," in *Proceedings of the 167th National Meeting of the American Chemical Society, Division of Fuel Chemistry*, Vol. 19, No. 1, 1974, pp. 111-122.
5. Fox, D. A., and White, A. H., "Effect of Sodium Carbonate upon Gasification of Carbon and Production of Producer Gas," *Ind. Eng. Chem.*, 23, 259 (1931).
6. Frank, F. H., and Meraikib, M., "Catalytic Influences of Alkalis on Carbon Gasification Reaction," *Carbon*, 8, 423-433 (1970).
7. Friedman, S., Hiteshue, R. W., and Schlisinger, M. D., "Hydrogenation of New Mexico Coal at Short Residence Time and High Temperature," *U.S. Bur. Mines, Rept. Invest.* (6470) (1964).
8. Gardner, N., "Catalysis in Coal Gasification," paper presented at the Symposium on Catalytic Conversion of Coal, Carnegie-Mellon Institute, April 21-23, 1975.
9. Gardner, N., Samuels, E., and Wilks, K., "Catalyzed Hydrogasification of Coal Chars," in *Proceedings of the 165th National Meeting of the American Chemical Society, Division of Fuel Chemistry*, Vol. 18, No. 2, 1973, pp. 43-85.
10. Haynes, W. P., Gasior, S. J., and Forney, A. J., "Catalysis of Coal Gasification at Elevated Pressures," in *Proceedings of the 165th National Meeting of the American Chemical Society, Division of Fuel Chemistry*, Vol. 18, No. 2, 1973, pp. 1-28.
11. Hill, G. R., "Project Western Coal: Conversion of Coal Into Liquids," Final Report to Office of Coal Research, Department of Interior, Contract No. 14-01-0001-271, May 1970.
12. Hiteshue, R. W., Friedman, S., and Madden, R., "Hydrogenation of Coal to Gaseous Hydrocarbons," *U.S. Bur. Mines, Rept. Invest.* (6027) (1962).
13. Hiteshue, R. W., Friedman, S. and Madden, R., "Hydrogasification of Bituminous Coals, Lignite, Anthracite, and Char," *U.S. Bur. Mines, Rept. Invest.* (6125) (1962).
14. Hoffman, E. J., "Behavior of Nickel Methanation Catalysts in Coal-Steam Reactions," in *Proceedings of the 163rd National Meeting of the American Chemical Society, Division of Fuel Chemistry*, Vol. 16, No. 2, 1972, 64-67.
15. Johnson, J. L., "Gasification of Montana Lignite in Hydrogen and in Helium During Initial Reaction Stages," paper to be presented at the National Meeting of the American Chemical Society, Division of Fuel Chemistry, Chicago, August 24-29, 1975.
16. Johnson, J. L., "Relationship Between the Gasification Reactivities of Coal Char and the Physical and Chemical Properties of Coal and Char," paper to be presented at the National Meeting of the American Chemical Society, Division of Fuel Chemistry, Chicago, August 24-29, 1975.
17. Katzman, H., "A Research and Development Program for Catalysis in Coal Conversion Processes," final report prepared for Electric Power Research Institute, Project EPRI 207-0-0, (May 1974).

18. Kawa, W., Friedman, S., Frank, L. V., and Hiteshue, R. W., "Coal Hydrogasification Catalyzed by Aluminum Chloride," in *Proceedings of the National Meeting of the American Chemical Society, Division of Fuel Chemistry*, Vol. 12, No. 2, 1968, pp. 43–47.
19. Kislykh, V. I., and Shishakov, N. V., "Catalytic Effect on Gasification in a Fluidized Bed," *Gaz. Prom.*, 8, 15–19 (1960).
20. Kroger, C., "The Gasification of Carbon With Air, Carbon Dioxide and Water Vapor and the Effect of Inorganic Catalysts," *Angew. Chem.*, 52, 129–139 (1939).
21. Kroger, C., and Melhorn, G., "The Behavior of Activated Bituminous Coal and Low-Temperature Carbonization and Gasification in a Current of Steam," *Brennst. Chem.*, 19, 257–262 (1938).
22. Long, F. J., and Sykes, K. W., "The Effect of Specific Catalysts on the Reactions of the Steam–Carbon System," *Proc. R. Soc.*, 215A, 100–10 (1952).
23. Matsura, K., Bodily, D. M., and Wiser, W. H., "Active Sites for Coal Hydrogenation," in *Proceedings of the 167th National Meeting of the American Chemical Society, Division of Fuel Chemistry*, Vol. 19, No. 1, 1974, pp. 157–162.
24. Mills, G. A., "Future Catalytic Requirements for Synthetic Energy Fuels," in *Proceedings of the 163rd National Meeting of the American Chemical Society, Division of Fuel Chemistry*, Vol. 16, No. 2, 1972, pp. 107–123.
25. Neumann, B., Kroger, C., and Fingas, E., "Die Wasserdampfzerzetzung und Kohlenstoff mit Aktivierenden Zusatzen," *Z. Anor. Allg. Chemie*, 197, 321–418 (1931).
26. Schroeder, W. G., Stevenson, L. G., and Stephenson, T. G., "Hydrogenation of Coal," U.S. Patent No. 3,152,063, October 1964.
27. Taylor, H. S., and Neville, H. A., "Catalysis in the Interaction of Carbon With Steam and With Carbon Dioxide," *J. Am. Chem. Soc.*, 43, 2055–2071 (1921).
28. Tomita, A., and Tamai, Y., "Hydrogenation of Carbons Catalyzed by Transition Metals," *J. Catal.*, 27, 293–300 (1972).
29. Vastola, F. J., and Walker, P. L., "The Reaction of Graphite 'Wear Dust' with Carbon Dioxide and Oxygen at Low Pressures," *J. Chem. Phys.*, 58, 20–24 (1961).
30. Vignon, L., "Water Gas," *Ann. Chim.*, 15, 42–60 (1921).
31. Walker, P. L., Shelef, M., and Anderson, R. A., "Catalysis of Carbon Gasification," in *The Chemistry and Physics of Carbon*, Vol. 4, Marcel Dekker, New York, 1968, pp. 287–383.
32. Weller, S., Clark, E. I., and Pelipetz, M. G., "Mechanism of Coal Hydrogenation," *Ind. Eng. Chem.*, 42, 334–336 (1950).
33. Weller, S., Pelipetz, M. G., Friedman, S., and Storch, H. H., "Coal Hydrogenation Catalysts-Batch Autoclave Tests," *Ind. Eng. Chem.*, 42, 330–334 (1950).
34. Wilks, K., Samuels, E., and Gardner, N., "The Gasification of Coal Chars," Master's thesis (of K. Wilks), Case Western Reserve University, 1974.

35. Willson, W. G. "Alkali Carbonate and Nickel Catalysis of Coal Steam Gasification." in *Proceedings of the 165th National Meeting of the American Chemical Society, Division of Fuel Chemistry*, Vol. 18, No. 2, 1973, pp. 29-42.
36. Wood, R. E., and Hill, G. R., "Coal Hydrogenation in Small Tube Reactors," in *Proceedings of the 163rd National Meeting of the American Chemical Society, Division of Fuel Chemistry*, Vol. 17, No. 1, 1972, pp. 28-36.

6 Kinetics of Bituminous Coal Char Gasification with Gases Containing Steam and Hydrogen

Quantitative correlations developed to describe coal char gasification kinetics are consistent with experimental data obtained under a wide range of conditions with both differential and integral contacting systems. The correlations are developed using data obtained at constant environmental conditions with a thermobalance apparatus and a differential fluid bed system at 1500–2000°F and 1–70 atm with a variety of gases and gas mixtures. The correlations are based on an idealized model of the gasification process which consists of three consecutively occurring stages: devolatilization, rapid-rate methane formation, and low-rate gasification.

Correlations to define quantitatively the effects of pertinent intensive variables on the kinetics of coal or coal char gasification reactions are necessary for the rational design of commercial systems to convert coal to pipeline gas. The available information which can be applied to the development of such correlations is relatively limited, particularly because the data reported from many studies conducted with integral contacting systems reflect, in part, undefined physical and chemical behavior peculiar to the specific experimental systems used. Although some differential rate data have been obtained with various carbonaceous materials, they cover only narrow ranges of the conditions potentially applicable to commercial gasification systems.

For the last several years the Institute of Gas Technology (IGT) has been conducting a study to obtain fundamental information on the gasification of coals and coal chars; this information could be used with selected literature information to develop engineering correlations which define quantitatively the effects of intensive variables on gasification rates over a wide range of conditions applicable to many gasification processes. The models and correlations developed at present are primarily applicable to bituminous coal chars prepared under mild or severe conditions in either inert or oxidizing atmospheres. Although we have achieved some success in applying these correlations to the gasification of subbituminous and lignite coals for limited conditions, the gasification kinetics of such materials have shown wide deviations from predictions of the correlations at lower temperatures and during initial stages of gasification.

The objectives of this paper are (a) to discuss the models which have been developed, (b) to present the correlations derived from these models, and (c) to demonstrate the consistencies between predictions of these correlations and various experimental gasification data obtained primarily with bituminous coal chars. The experimental information used to develop the models came from two main sources. Initial development of the model applied to the gasification of devolatilized coal char in hydrogen and steam-hydrogen mixtures was based both on data from an IGT study with a high pressure thermobalance apparatus and on differential rate data from the Consolidation Coal Co. on the gasification of Disco char in a small-scale fluid bed [1-3]. The model was extended to the gasification of char containing volatile matter and to gasification with gases containing carbon monoxide, carbon dioxide, and methane as well as steam and hydrogen using data primarily from the thermobalance study.

The thermobalance is particularly useful in obtaining fundamental gasification information because gasification rates can be measured at constant, well defined environmental conditions with it. Most of the information used to formulate the kinetic models developed was based on data from several hundred tests conducted with the thermobalance; this apparatus is described below.

EXPERIMENTAL

The thermobalance is an apparatus capable of continuously weighing a coal sample which is undergoing reaction in a gaseous environment of desired composition at a constant pressure. The temperature can be kept constant or varied ($10°$F/min is the maximum rate for the apparatus used at IGT). The nature of gas-solid contact with the apparatus used in this study is shown in Figure 1. The coal sample is contained in the annular space of a wire mesh basket bounded on the inside by a hollow, stainless steel tube and on the outside by a wire mesh screen. To facilitate mass and heat transfer between the bed and its environment, the thickness of the bed is only 2-3 particle diameters when using $-20 + 40$ US sieve-size particles. Gas flow rates with this sytem are sufficiently large relative to gasification rates, so that gas conversion is limited to less than 1% for devolatilized coal char.

In a typical test the wire mesh basket is initially in an upper, cooled portion of the reactor in which a downward, inert gas flow is maintained. During this time the desired temperature and pressure conditions are established in a lower, heated portion of the reactor in the presence of a flowing gas. A test is initiated by lowering the basket into the heated reaction zone, a procedure which takes 5-6 sec. Theoretical computation shows that about 2 min are needed for the sample to achieve reactor temperature as measured by several thermocouples surrounding the basket in the reaction zone. This computation is reasonably corroborated by various kinetic indications and by the behavior of the thermocouples in re-attaining their preset temperatures. The sample is kept in the heated portion of the reactor for the specified time while its weight is continuously recorded. The test is terminated by raising the basket back to the upper, cooled portion of the reactor.

During a test, the dry feed gas flow rates are measured by an orifice meter and the dry product gas flow rates measured by a wet-test meter. Periodic samples of product gas are taken to determine the composition by mass spec-

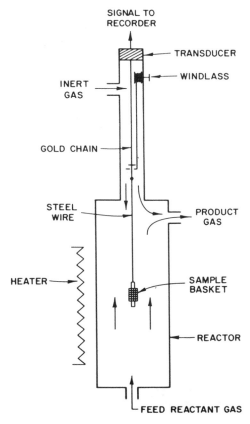

Figure 1 Thermobalance reactor.

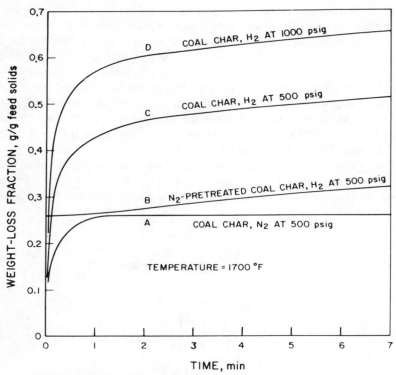

Figure 2 Typical weight-loss curves obtained with high pressure thermobalance for gasification in hydrogen and in nitrogen.

TABLE I. COMPOSITION OF AIR-PRETREATED hvab PITTSBURGH NO. 8 COAL CHAR (IRELAND MINE)

Ultimate Analysis, dry	wt %
Carbon	71.1
Hydrogen	4.26
Oxygen	8.85
Nitrogen	1.26
Sulfur	3.64
Ash	10.89
Total	100.00
Proximate Analysis, dry	
Fixed carbon	60.7
Volatile	28.4
Ash	10.9
Total	100.0

trometer. Feed and product steam flow rates are measured gravimetrically, and the solids residues are analyzed for total carbon and hydrogen.

Figure 2 shows typical, smoothed, weight loss-vs.-time characteristics obtained using an air-pretreated hvab Pittsburgh coal char from the Ireland mine. These curves are discussed in more detail later. The composition of the coal char, used extensively in the experimental study, is given in Table I.

KINETIC MODELS

When a coal or coal char containing volatile matter is initially subjected to an elevated temperature, a series of complex physical and chemical changes occur in coal's structure, accompanied by thermal pyrolysis reactions which result in devolatilization of certain coal components. The distribution of the evolved products of the reactions, which initiate at less than 700°F and can be considered to occur almost instantaneously at temperatures greater than 1300°F, is generally a function of the temperature, pressure, and gas composition existing during devolatilization and of the subsequent thermal and environmental history of the gaseous phase (including entrained liquids) prior to quenching.

When devolatilization occurs in the presence of a gas containing hydrogen at an elevated pressure, in addition to thermal pyrolysis reactions, coals or coal chars containing volatile matter also exhibit a high although transient reactivity for methane formation. Although some investigators have suggested that this process occurs simultaneously with thermal pyrolysis reactions, studies done with a greater time resolution indicated that this rapid-rate methane formation occurs at a rate which is at least an order of magnitude slower than devolatilization [4, 5]. In this sense it occurs after devolatilization.

The amount of carbon gasified to methane during the transient high reactivity increases significantly with increases in hydrogen partial pressure [4–6]. Experimental evidence indicates that at sufficiently high hydrogen partial pressures virtually all of the carbon not evolved during devolatilization can be gasified quickly to methane by this process [6]. This is contrary to some proposed models which assume that only a limited amount of carbon can be gasified in this reaction stage regardless of the hydrogen partial pressure [7, 8].

At temperatures greater than 1700°F the transient reactivity for rapid rate methane formation exists only briefly. For coals or coal chars prepared in inert atmospheres this period is seconds or less [6]. IGT's studies suggest that for air-pretreated coal chars, this period is more extended although the total amounts of carbon which can be gasified by this process at a given temperature and hydrogen partial pressure are comparable for coals and coal chars prepared at sufficiently low temperatures in either inert gas or air.

After the devolatilization and rapid-rate methane formation stages are completed, char gasification occurs at a relatively slow rate; various models to describe the gasification kinetics of this material for various limited ranges of conditions have been proposed. The differential rates of reaction of devolatilized coal chars are a function of temperature, pressure, gas composition, carbon conversion, and prior history.

General Assumptions in the Development of Models

The models developed in this study for the quantitative description of coal char gasification kinetics assume that the overall gasification occurs in three consecutive stages: (1) devolatilization, (2) rapid-rate methane formation, and (3) low-rate gasification. The reactions in these stages are independent. Further, a feed coal char contains two types of carbon—volatile carbon and base carbon. Volatile carbon can be evolved solely by thermal pyrolysis, independent of the gaseous medium. The distribution of evolved products derived from the volatile carbon fraction is not defined in the model. In any application of the model to an integral contacting system, the devolatilization products are estimated by extrapolation or interpolation of yields obtained in pilot-scale fluid or moving-bed systems. This procedure can be used for narrow ranges of conditions and only for very similar contacting systems. Base carbon remains in the coal char after devolatilization is complete. This carbon can be subsequently gasified in either the rapid-rate methane formation stage or the low-rate gasification stage.

Initial amounts of volatile and base carbon are estimated from standard analysis of the feed coal char:

C_v (volatile carbon), grams/gram feed coal

$$= C_t^0 \text{ (total carbon), grams/gram feed coal}$$
$$- C_b^0 \text{ (base carbon), grams/gram feed coal} \quad [1]$$

where C_t^0 represents the total carbon in the feed coal char obtained from an ultimate analysis, and C_b^0 represents the carbon in the fixed carbon fraction of the feed coal as determined in a proximate analysis. Note that C_b^0 does not equal the fixed carbon fraction because the fixed carbon fraction includes, in addition to carbon, other organic coal components not evolved during standard devolatilization.

Experimental results from thermobalance or free-fall tests conducted at IGT indicate that the assumption of a constant volatile carbon fraction is valid for coal heat-up rates as high as 200°F/sec. However, other studies conducted with extremely rapid heat-up rates (10^4 to 10^7°F/sec) have shown that quantities of carbon evolved during thermal pyrolysis can exceed the volatile carbon fraction defined in this model [9, 10]. An allowance for the increase in evolved carbon would, therefore, have to be made in systems using such high heating rates.

The base-carbon conversion fraction, X, is defined as:

$$X = \frac{\text{base carbon gasified}}{\text{base carbon in feed coal char}} = \frac{C_b^0 - C_b}{C_b^0} \quad [2]$$

where C_b = base carbon in coal char at an intermediate level of gasification, grams/gram feed coal char.

When making a kinetic analysis of the thermobalance data it was necessary to relate the measured values of weight-loss fraction, $\Delta W/W_0$, to the base carbon conversion fraction, X. When devolatilization is complete, essentially all of the organic oxygen has been gasified. Thereafter, additional weight loss, which primarily results from gasification of the base carbon, is accompanied by the evolution of a constant fraction of noncarbon components in the coal such as nitrogen, hydrogen, and sulfur. Estimates of an average value of the fraction of noncarbon components gasified along with the base carbon for each type of coal char tested have been based on analyses of char residues obtained in thermobalance and pilot-scale fluid-bed tests. With this simplifying assumption, the following relationships result:

$$C_b^0 = (1 - V - A)(1 - \gamma) \quad [3]$$

and

$$C_b^0 = C_b - [(\Delta W/W_0) - V](1 - \gamma) \quad [4]$$

where

V = volatile matter in feed-coal char (including moisture), grams/gram feed coal char

A = nongasifiable matter in feed coal char (including ash and some sulfur), grams/gram feed coal char

γ = noncarbon matter evolved along with base carbon, grams/gram base carbon evolved

Thus, from Equations 2, 3, and 4

$$X = \frac{(\Delta W/W_0) - V}{1 - V - A} \quad [5]$$

The results in Figure 2 can be interpreted by the three reaction stages defined above. Curve A corresponds to a test in which air-pretreated Ireland mine coal char was exposed to a nitrogen atmosphere at 500 psig. During the first minute when the sample is heating in the thermobalance, the weight loss corresponds to the evolution of volatile matter. After this period no further significant weight loss occurs. The total weight loss of ca. 26% corresponds closely to the volatile matter in the feed coal char obtained by standard proximate analysis. Within the context of the three reaction stages defined, weight loss in this text occurs entirely in the devolatilization stage where all of the volatile carbon has been gasified; all of the base carbon remains in the devolatilized coal char. When the char resulting from this text is then exposed to hydrogen at 500 psig (curve B), the reaction of the hydrogen with base carbon to form methane results in further weight loss. This reaction, which takes place at a much slower rate than devolatilization, occurs in the low-rate gasification stage. With this particular sample there was no reaction in the rapid-rate methane formation stage because the reactivity of the coal char in this stage was destroyed during prolonged exposure to nitrogen.

Weight loss as a result of reaction during rapid-rate methane formation is illustrated by curve C where a sample of the original coal char was exposed only to hydrogen at 500 psig with no initial exposure to nitrogen. The weight loss during the first minute of this test is considerably greater than the corresponding weight loss during this period when the coal char was exposed to nitrogen (curve A). The difference in the weight loss between curves C and A during the first minute is caused by gasification of base carbon in the rapid-rate methane formation stage. Weight losses of the amount shown by curve B during this initial period are negligible. This is consistent with the assumption that base carbon gasification in the rapid-rate methane formation stage and in the low-rate gasification stage occurs consecutively. Curve D is qualitatively similar to curve C except that there is a greater weight loss from rapid-rate methane formation resulting from the higher hydrogen pressure.

Correlations for Rapid-Rate Methane Formation Stage

The amount of base carbon gasified during the rapid-rate methane formation stage can be estimated by the base-carbon conversion level, X_R, obtained from weight loss-vs.-time characteristics 2 min after a sample is lowered into the reactor. As indicated previously, this corresponds to the time required for coal heat up. During this period negligible conversion occurs in the low-rate gasification stage although devolatilization and rapid-rate methane formation reactions should be complete at temperatures above ca. 1500°F. Values of X_R have been correlated with hydrogen partial pressure, P_{H_2}, according to the following expression for data obtained in tests conducted at varied conditions:

$$M(X_R) = \int_0^{X_R} \frac{\exp(+\alpha X^2)\,dX}{(1-X)^{2/3}]} = 0.0092 f_R P_{H_2} \qquad [6]$$

where

P_{H_2} = hydrogen partial pressure, atm

f_R = relative reactivity factor for rapid-rate methane formation dependent on the particular carbonaceous solid (defined as unity for air-pretreatment Ireland mine coal char)

α = kinetic parameter dependent on gas composition and pressure

The value of α in the above expression is ca. 0.97 for tests done in pure hydrogen or in hydrogen–methane mixtures and is approximately equal to 1.7 for a variety of gas compositions containing steam and hydrogen. This parameter is discussed in greater detail in a later section of the low-rate gasification stage. However, for the case where $\alpha = 0.97$, then $M(X_R) \cong -\ln(1 - X_R)$.

Figure 3 is a plot of the function $M(X_R)$ vs. hydrogen partial pressure, P_{H_2}, for tests conducted on the thermobalance with air-pretreated Ireland mine coal char; note the good agreement between these data and the correlation form

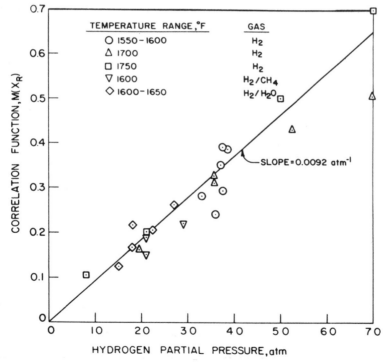

Figure 3 Correlation of base carbon conversion for gasification in the rapid-rate methane formation stage (thermobalance data).

given in Equation 6. Equation 6 also appears to be reasonably applicable to the gasification of some coals as well as to coal chars. Data obtained by Birch et al. [11] for the hydrogenation of Brown coal in a fluid bed and by Hiteshue et al. [12] for the hydrogenation of a hvab Pittsburgh coal in a fixed bed are given in Figures 4 and 5. The procedures used to treat the data from these two investigations have been described [13]. Relatively small variations in f_R values are exhibited by the different materials. Similar small degrees of variation have also been noted for other bituminous coal chars tested at IGT using the thermobalance.

The gasification of base carbon in the rapid-rate methane formation stage depends apparently only on hydrogen partial pressure and not on the partial pressures of other gaseous species normally present in gasifying atmospheres. This is partly indicated in Figure 3 for tests conducted with hydrogen–methane and hydrogen–steam mixtures and has also been observed with synthesis-gas mixtures. In a system containing no hydrogen, base carbon is not evolved except through gasification in the low-rate gasification stage. In Figure 6 this effect is illustrated for tests conducted experimental apparatus (approximately 30°F/sec). The fact that this correlation seems to apply for data obtained in other experimental systems where heat-up rates as high as 200°F/sec were used suggests

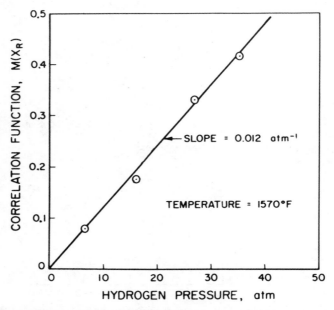

Figure 4 Correlation of base carbon conversion for gasification in a fluid bed in the rapid-rate methane formation stage.

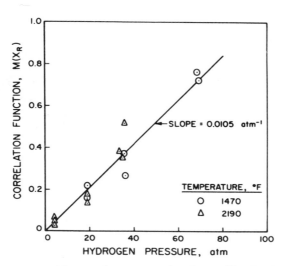

Figure 5 Correlation of base carbon conversion for gasification in a fixed bed in the rapid-rate methane formation stage.

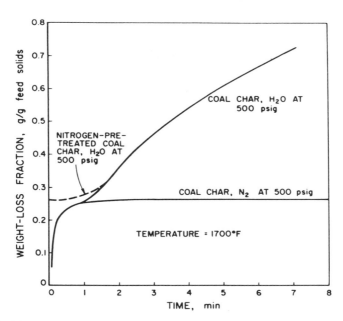

Figure 6 Weight loss curves obtained in high pressure thermobalance for gasification in steam and in nitrogen.

that within a limited range of heat-up rates, base carbon conversion in the rapid-rate methane formation stage is independent of heat-up rate for final temperatures greater than 1500°F. This conclusion applies only when reaction in the rapid-rate methane formation stage goes to completion. At sufficiently low temperatures, the amount of base carbon conversion which can be attributed to rapid-rate methane formation is less than that which would be predicted by Equation 6, even for exposure to hydrogen for as long as 1 hr [13].

Data obtained in experiments done at low temperatures, such as those in Figure 7, have been correlated using a more detailed model to describe the rapid-rate methane formation process prior to the completion of this reaction. This model is described in a previous publication [13]. Certain characteristics of this model rationalize the independence of total base carbon conversion in the rapid-rate methane formation stage from heating rate and final temperature for tests done above 1500°F. The following critical steps assumed in the model relate to this range of conditions:

$$A_0 \longrightarrow A_* \qquad [7]$$

$$A_* \longrightarrow B \qquad [8]$$

$$\text{base carbon + hydrogen} \xrightarrow{A_*} \text{methane} \qquad [9]$$

This model, which is qualitatively similar to one proposed by Mosely and Patterson [6], assumes that the coal char initially forms an active intermediate A_*, (Equation 7) which catalyzes the reaction between base carbon and gaseous hydrogen to form methane (Equation 9). This reaction, however, competes with a reaction in which A_* deactivates to form the inactive species, B (Equation 8).

The following expression is assumed to describe the rate of reaction in Equation 9:

$$dX/dt = f_R k_3(T) P_{H_2} (1 - X)^{2/3} \exp(-\alpha X^2) N_{A_*} \qquad [10]$$

where

$k_3(T)$ = rate constant dependent on temperature, T

N_{A_*} = concentration of species, A, mole/mole of base carbon

t = time

The dependence of the conversion rate, dX/dt, on conversion fraction, X, shown in Equation 10 is the same as that used in correlations presented in a later section which were developed to describe gasification in the low-rate gasification stage. With the models assumed, the term $(1 - X)^{2/3}$ is proportional

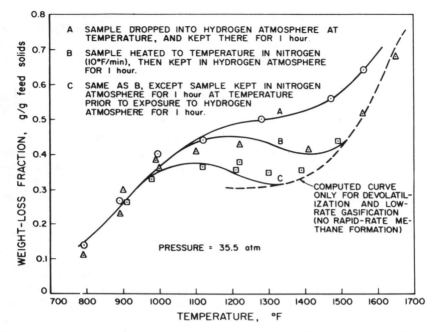

Figure 7 Effect of prior exposure to nitrogen on gasification in the rapid-rate methane formation stage.

to the effective surface area undergoing gasification, and the term $\exp(-\alpha X^2)$ represents the relative reactivity of the effective surface area which decreases with increasing conversion level for positive values of α.

The reaction rates in Equations 7 and 8 are assumed to be first order, leading to the expression:

$$d(N_{A_*} + N_{A_0})/dt = -k_2(T) \cdot N_{A_*} \quad [11]$$

where

$k_2(T)$ = rate constant, dependent on temperature

N_{A_0} = concentration of species, A_0, mole/mole base carbon

Combining Equations 10 and 11 leads to:

$$\frac{dX}{d(N_{A_*} + N_{A_0})} = -f_R \frac{k_3(T)}{k_2(T)} P_{H_2}(1-X)^{2/3} \exp(-\alpha X^2) \quad [12]$$

It it is assumed that $k_3(T)/k_2(T)$ is independent of temperature and is equal to β, Equation 12 can be integrated to yield the following expression for the condition at which all of A_0 has been converted to B.

$$M(X_R) = \int_0^{X_R} \frac{\exp(\alpha X^2)\, dX}{(1-X)^{2/3}]} = f_R \beta \cdot N_{A_0}^0 P_{H_2} \qquad [13]$$

A comparison of this expression with Equation 6 indicates that $\beta N_{A_0}^0 = 0.0092$ atm^{-1}. Since no definition of the temperature history was required to develop Equation 13, the suggested model indicates that the amount of base carbon conversion to methane during the rapid-rate methane formation step is independent of heat-up rate or temperature level when the intermediate, A, has been completely deactivated.

Correlations for Low-Rate Gasification Stage

For practical purposes coal chars undergo low-rate gasification only after the devolatilization and rapid-rate methane formation reactions are completed. Results obtained with the thermobalance indicate that at greater than 1500°F char reactivity over a major range of carbon conversion in the low-rate stage is substantially the same whether devolatilization occurs in nitrogen or in a gasifying atmosphere under the same conditions. Therefore, this study treats low-rate char gasification as a process essentially independent of devolatilization conditions with one important exception, the temperature of devolatilization. It has been shown in this study as well as by Blackwood [14] that the reactivity of char at a given temperature, T, decreases with increasing pretreatment temperature, T_0, when $T_0 > T$. This effect is quantitatively represented in the correlations below. However, the model adopted does not account for pretreatment effects on gasification during initial stages of char gasification which occur particularly at gasification temperatures less than 1600°F. At these lower temperatures, specific pretreatment conditions such as gas atmosphere and time of pretreatment produce complex effects during subsequent gasification for base carbon conversions of less than 10% [15]. These limitations have no practical significance in using the simplified model developed to describe coal char gasification kinetics at higher temperatures or for base-carbon conversion levels sufficiently greater than 10%.

The gasification data of Zielke and Gorin [2, 3] and Goring et al. [1] for fluid-bed gasification of Disco char as well as the bulk of data obtained in IGT studies with the high pressure thermobalance and pilot-scale fluid beds were used to evaluate parameters in a quantitative model developed to describe coal-char-gasification kinetics over a wide range of conditions in the low-rate gasification stage. Three basic reactions were assumed to occur in gases containing steam and hydrogen:

Reaction I: $H_2O + C \rightleftharpoons CO + H_2$

Reaction II: $2H_2 + C \rightleftharpoons CH_4$

Reaction III: $H_2 + H_2O + 2C \rightleftharpoons CO + CH_4$

Reaction I is the conventional steam–carbon reaction, which is the only one that occurs in pure steam at elevated pressures or with gases containing steam at low pressure. Although at elevated temperatures this reaction is affected by thermodynamic reversibility only for relatively high steam conversions, the reaction is severely inhibited by the poisoning effects of hydrogen and carbon monoxide at steam conversions far removed from equilibrium for this reaction. Some investigators have also noted inhibition by methane. (Although some methane has been detected in gaseous reaction products when gasification is conducted with pure steam, it is uncertain whether this methane results from the direct reaction of steam with carbon or from the secondary reaction of hydrogen, produced from the steam–carbon reaction, with the carbon in the char.)

Reaction II, the only reaction that could occur in pure hydrogen or in hydrogen–methane mixtures, depends greatly on the hydrogen partial pressure. Many investigators have found that at elevated pressures its rate is directly proportional to the hydrogen partial pressure.

The stoichiometry of Reaction III limits its occurrence to systems in which both steam and hydrogen are present. Although this reaction is the stoichiometric sum of Reactions I and II, this model considers it to be a third, independent gasification reaction. Reaction III, arbitrarily assumed to occur in the development of this model to facilitate correlation of experimental data, has been suggested by Blackwood and McGrory [16] as being necessary in such a system. Curran and Gorin [17] also assumed this reaction to correlate kinetic data for gasification of lignite at 1500°F in steam–hydrogen-containing gases.

The correlations developed in this study to describe kinetics in the low-rate gasification stage are summarized as follows:

$$dX/dt = f_L k_T (1 - X)^{2/3} \exp(-\alpha X^2) \quad [14]$$

where

$$k_T = k_I + k_{II} + k_{III} \quad [15]$$

Here, k_I, k_{II}, and k_{III} are rate constants for the individual reactions considered. It is assumed that each of the three reactions occurs independently but that the rate of each is proportional to the same surface area term, $(1 - X)^{2/3}$ and surface reactivity term, $\exp(-\alpha X^2)$.

Individual parameters in Equations 14 and 15 are defined as functions of temperature and pressure according to:

$$f_L = f_0 \exp(8467/T_0) \qquad [16]$$

$$k_I = \frac{\exp(9.0201 - 31{,}705/T)\left(1 - \dfrac{P_{CO} P_{H_2}}{P_{H_2O} K_I^E}\right)}{\left[1 + \exp(-22.2160 + 44{,}787/T)\left(\dfrac{1}{P_{H_2O}} + 16.35 \dfrac{P_{H_2}}{P_{H_2O}} + 43.5 \dfrac{P_{CO}}{P_{H_2O}}\right)\right]^2} \qquad [17]$$

$$k_{II} = \frac{P_{H_2}^2 \exp(2.6741 - 33{,}076/T)\left(1 - \dfrac{P_{CH_4}}{P_{H_2}^2 K_{II}^E}\right)}{[1 + P_{H_2} \exp(-10.4520 + 19{,}976/T)]} \qquad [18]$$

$$k_{III} = \frac{P_{H_2}^{1/2} P_{H_2O} \exp(12.4463 - 44{,}544/T)\left(1 - \dfrac{P_{CH_4} P_{CO}}{P_{H_2} P_{H_2O} K_{III}^E}\right)}{\left[1 + \exp(-6.6696 + 15{,}198/T)\left(P_{H_2}^{1/2} + 0.85 P_{CO} + 18.62 \dfrac{P_{CH_4}}{P_{H_2}}\right)\right]^2} \qquad [19]$$

$$\alpha = \frac{52.7 P_{H_2}}{1 + 54.3 P_{H_2}} + \frac{0.521 P_{H_2}^{1/2} P_{H_2O}}{1 + 0.707 P_{H_2O} + 0.50 P_{H_2}^{1/2} P_{H_2O}} \qquad [20]$$

where

$K_I^E, K_{II}^E, K_{III}^E$ = equilibrium constants for Reactions I, II, and III, considering carbon as graphite

T = reaction temperature, °R

T_0 = maximum temperature to which char has been exposed prior to gasification, °R (if $T_0 < T$, then a value of $T_0 = T$ is used in Equation 16)

$P_{H_2}, P_{H_2O}, P_{CO}, P_{CH_4}$ = partial pressures of H_2, H_2O, CO, and CH_4, atm

f_0 = relative reactivity factor for low-rate gasification which depends on the particular carbonaceous solid

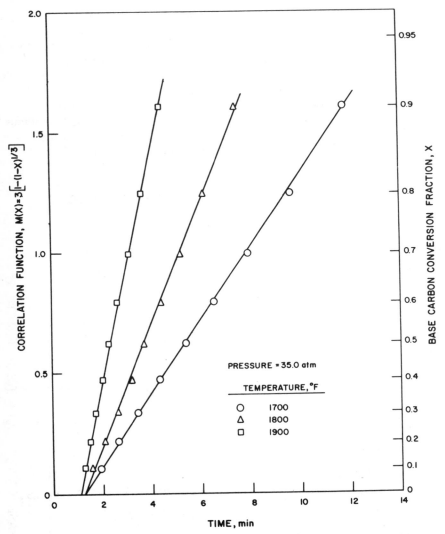

Figure 8 Effect of temperature on low-rate gasification in steam (thermobalance data).

Values of f_0 obtained in this study were based on the definition $f_0 = 1$ for a specific batch of air-pretreated Ireland mine coal char. Samples of this coal char obtained from different air-pretreatment tests exhibited some variations in reactivity as determined by thermobalance tests conducted at standard conditions. The values of f_0 so determined ranged from approximately 0.88 to 1.05. Results of tests made with the thermobalance, using a variety of coals and coal chars, have indicated that the relative reactivity factor, f_0, generally tends to increase with decreasing rank although individual exceptions to this trend exist. Values have been obtained which range from 0.3 for a low-volatile bituminous coal char to about 10 for a North Dakota lignite. The reactivity of the Disco char used in gasification studies conducted by the Consolidation Coal Co. [1, 2, 3] is $f_0 = 0.488$.

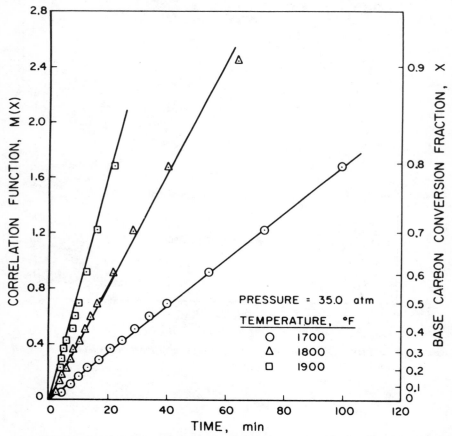

Figure 9 Effect of temperature on low-rate gasification in hydrogen (thermobalance data).

KINETIC MODELS 301

An integrated form of Equation 14 was used to evaluate certain parameters in the above correlations, based on data obtained with the thermobalance.

$$M(X) = \int_0^X \frac{\exp(+\alpha X^2)\,dX}{(1-X)^{2/3}} = \int_0^{X_R} \frac{\exp(+\alpha X^2)\,dX}{(1-X)^{2/3}}$$

$$+ \int_{X_R}^X \frac{\exp(-\alpha X^2)\,dX}{(1-X)^{2/3}} = \int_0^{X_R} \frac{\exp(+\alpha X^2)\,dX}{(1-X)^{2/3}} + f_L k_T t \quad [21]$$

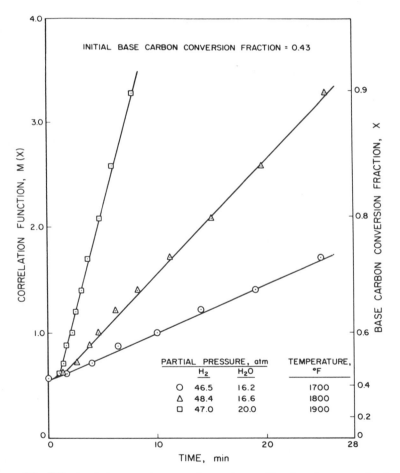

Figure 10 Effect of temperature on low-rate gasification in steam–hydrogen mixtures (thermobalance data).

The term

$$\int_0^{X_R} \frac{\exp(+\alpha X^2)\,dX}{(1-X)^{2/3}}$$

was evaluated from Equation 6 for tests in which no nitrogen pretreatment was used. For tests in which the feed coal char was initially devolatilized in nitrogen at the same temperature and pressure to be subsequently used for gasification in a gasifying atmosphere, reactivity for rapid-rate methane formation was destroyed above 1500°F and $X_R = 0$. For this condition

$$M(X) = \int_0^X \frac{\exp(+\alpha X^2)\,dX}{(1-X)^{2/3}]} = f_L k_T t \qquad [22]$$

Typical plots of $M(X)$ vs. t are given in Figures 8–10 for data obtained with air-pretreated Ireland mine coal char. The slopes of the lines drawn correspond

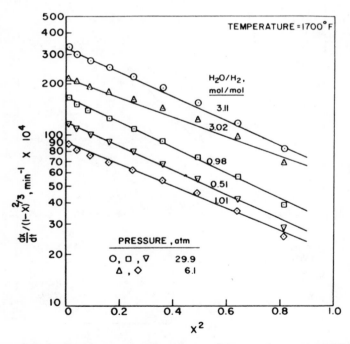

Figure 11 Effects of pressure and gas composition on low-rate gasification in steam–hydrogen mixtures [1, 3].

to values of $f_L k_T$ characteristic of each test. Results of tests conducted with pure steam (Figure 8) were correlated using $\alpha = 0$ and were consistent with Equation 20, which corresponds to the situation in which specific gasification rates, $[(dX/dt)/(1 - X)]$, increase with an increasing carbon conversion level. For gasification in hydrogen or steam-hydrogen mixtures (Figures 9 and 10), however, specific gasification rates generally decrease with increasing conversion level.

Carbon gasification rates were directly measured in the fluid-bed tests conducted with Disco char [1, 2, 3] and values of $f_L k_T$ and α can be obtained graphically by plotting values of $\ln [(dX/dt)/(1 - X)^{2/3}]$ vs. X^2. Such a plot is illustrated in Figure 11, where $[(dX/dt)/(1 - X)^{2/3}]$ at $X = 0$ is equal to $f_L k_T$, and the slope of a given line is equal to α. Generally the correlations presented are consistent with values of $f_L k_T$ and α obtained for the gasification of Disco char at 1600°F and 1700°F and 1, 6, and 30 atm for gasification with pure

TABLE II. EXPERIMENTAL AND CALCULATED RATE CONSTANTS FOR GASIFICATION OF AIR-PRETREATED IRELAND MINE COAL CHAR IN HYDROGEN

Temp, °F	H_2 Pressure, atm	Base Carbon Conversion Range, Fraction	Rate Constant, fLT, min^{-1}	
			Experimental	Calculated
1600	36.5	0–0.314	0.0068	0.0069
1650	36.3	0–0.541	0.0117	0.0112
1650	36.7	0–0.592	0.0111	0.0113
1650	36.7	0–0.654	0.0097	0.0113
1700	19.3	0–0.197	0.0106	0.0090
1700	36.2	0–0.329	0.0167	0.0181
1700	53.2	0–0.449	0.0276	0.0273
1700	69.9	0–0.508	0.0292	0.0365
1700*	35.1	0–0.802	0.0175	0.0175
1750	35.4	0–0.348	0.0261	0.0277
1750	35.4	0–0.509	0.0274	0.0277
1750	36.1	0–0.640	0.0237	0.0283
1750	35.2	0–0.367	0.0272	0.0276
1770*	1.0	0	0	0.0002
1770*	18.1	0–0.069	0.0175	0.0152
1770*	36.6	0–0.134	0.0350	0.0341
1770*	36.1	0–0.570	0.0290	0.0335
1770*	52.9	0–0.180	0.0500	0.0510
1770*	69.9	0–0.250	0.0700	0.0686
1800	21.6	0–0.263	0.0222	0.0239
1800	49.9	0–0.263	0.0641	0.0617
1800*	36.0	0–0.910	0.0416	0.0414
1900*	35.3	0–0.850	0.0910	0.0920

* In these tests, the air-pretreated coal char was either initially devolatilized in nitrogen for 1 hr at the temperature subsequently used for gasification in a gasifying medium or it was devolatilized and partially gasified in an integral fluid-bed test using a steam-hydrogen feed gas.

hydrogen and with steam-hydrogen mixtures. The correlations are also consistent with individual rates of methane and carbon oxide formation reported in the studies done with Disco char for initial levels fo carbon conversion. Although the relative rates of methane-to-carbon oxide formation were reported to increase somewhat with increasing conversion level (a trend not accounted for in the model developed in this study), investigators at the Consolidation Coal Co. have suggested that this effect was caused by a catalytic reaction downstream of the fluid-bed reactors used, in which some of the carbon monoxide produced in the reactor was converted to methane [17].

The consistency between the calculated and experimental values of $f_L k_T$ for tests conducted under various conditions with the thermobalance using air-

TABLE III. EXPERIMENTAL AND CALCULATED RATE CONSTANTS FOR GASIFICATION OF AIR-PRETREATED IRELAND MINE COAL CHAR IN STEAM AND STEAM-HYDROGEN MIXTURES

Temp, °F	Partial Pressure, atm		Base Carbon Conversion Range, Fraction	Rate Constant, $f_L k_T$, min^{-1}	
	H_2	H_2O		Experimental	Calculated
1700	—	35.0	0–0.941	0.1547	0.1736
1700	27.3	29.8	0–0.977	0.0598	0.0528
1700	12.8	23.6	0–0.327	0.0706	0.0566
1700	17.8	18.4	0–0.272	0.0365	0.0385
1700	19.2	17.5	0–0.651	0.0356	0.0361
1700	18.2	17.9	0–0.816	0.0344	0.0373
1700	23.3	12.4	0–0.300	0.0339	0.0274
1700	23.2	12.5	0–0.429	0.0262	0.0275
1700	23.8	11.9	0–0.511	0.0190	0.0267
1700	25.0	11.3	0–0.725	0.0224	0.0261
1700[a]	13.7	48.5	0.43–0.908	0.0831	0.1110
1700[a]	46.5	16.2	0.43–0.781	0.0446	0.0385
1700[a]	28.3	32.5	0.43–0.822	0.0425	0.0564
1750	17.6	18.5	0–0.801	0.0526	0.0763
1750	18.3	18.2	0–0.495	0.0613	0.0741
1750	17.6	18.5	0–0.702	0.0667	0.0763
1750	18.0	18.5	0–0.940	0.0917	0.0758
1750	5.3	30.8	0–0.573	0.1803	0.2162
1750	23.2	13.2	0–0.320	0.0441	0.0532
1750	26.9	9.2	0–0.334	0.0406	0.0422
1750	17.6	18.3	0–0.350	0.0670	0.0815
1800	—	35.0	0–0.990	0.2559	0.2815
1800	48.4	16.6	0–0.956	0.1115	0.1130
1800	32.4	27.6	0–0.981	0.1643	0.1710
1900	—	35.0	0–0.992	0.4795	0.4366
1900[a]	17.2	42.7	0.43–0.996	0.6570	0.8750
1900[a]	47.0	20.0	0.43–0.999	0.4000	0.3900
1900[a]	33.2	33.2	0.43–0.997	0.680	0.607

[a] See Table II.

pretreated Ireland mine coal char is demonstrated in Tables II–IV and Figures 12-14. The correlations developed have also been used to predict behavior in pilot-scale, moving- and fluid-bed tests conducted at IGT and elsewhere. The assumptions made in characterizing the nature of gas–solids contacting in these integral systems have been described previously [13]. The most important assumptions made for fluid-bed systems are (a) the gas in the fluid bed is perfectly mixed, and (b) when continuous solids flow is used, the solids are in plug flow. For moving-bed systems we assumed that both gas and solids were in plug flow. With these simplifying assumptions, the conditions of primary importance in characterizing integral gasification behavior in individual tests include coal char feed rate and composition, particle residence times in the reactor, reactor temperature, pressure, feed gas composition, and flow rate. When coal char containing volatile matter was used as a feed material, rapid-rate methane formation and devolatilization were assumed to occur in a free-fall space above the reaction beds used. When devolatilized coal char was the feed material, no rapid-rate methane formation was considered to occur. Predicted and experimental integral

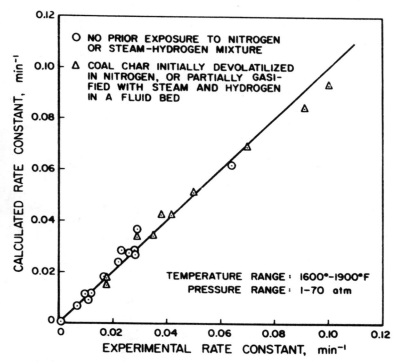

Figure 12 Experimental and calculated rate constants for low-rate gasification in hydrogen (thermobalance data).

TABLE IV. EXPERIMENTAL AND CALCULATED RATE CONSTANTS FOR GASIFICATION OF AIR-PRETREATED IRELAND MINE COAL CHAR IN SYNTHESIS GAS

Temp, °F	Partial Pressure, atm				
	CO	CO_2	H_2	H_2O	CH_4
1550	0.17	0.32	5.92	28.11	1.68
1600	0.95	0.28	14.46	11.07	10.73
1600	0.65	0.35	12.74	13.82	9.68
1600	0.65	0.38	14.49	11.33	10.29
1600	0.17	0.17	7.73	14.66	9.65
1600	0.10	0.33	6.41	26.58	2.18
1600	8.60	6.38	11.07	7.50	2.38
1600	7.57	1.70	18.87	6.84	1.04
1600	2.28	0.46	19.03	6.57	8.48
1650	0.55	0.26	11.76	14.66	9.65
1650	0.10	0.29	6.27	26.47	2.51
1650	2.24	3.25	6.00	23.02	1.29
1700	1.85	1.23	8.47	24.71	0.04
1700	6.05	1.64	15.81	12.91	0.16
1700	5.27	4.17	10.43	15.87	0.05
1700	4.52	3.79	9.54	18.11	0.12
1700	6.90	4.45	10.88	13.58	0.14
1700	2.26	3.12	3.49	28.07	0.11
1700	4.33	4.69	4.18	22.54	0.07
1700ª	0.60	2.80	4.30	60.10	0.40
1700ª	5.0	9.4	19.8	25.8	2.1
1700ª	1.8	3.1	6.3	19.2	0.4
1700ª	5.9	10.8	13.2	28.0	0.9
1700ª	4.3	5.1	6.7	16.8	0.5
1750	6.07	4.62	9.69	13.47	2.34
1750	3.31	4.80	8.12	19.08	1.23
1750	2.49	3.94	4.99	24.18	0.94
1750	5.87	6.61	7.03	16.00	0.16
1750	6.6	4.9	10.6	13.9	0.3
1800ª	6.2	7.4	18.1	31.6	0.3
1800ª	1.8	4.9	11.6	47.7	0.2
1800ª	10.7	6.7	17.7	19.1	1.5
1800ª	4.5	2.6	7.4	13.3	0.3
1800ª	6.9	3.4	10.7	8.4	0.4
1800ª	6.0	9.7	13.4	30.5	0.4
1800ª	2.9	11.1	14.1	36.5	0.4
1800ª	5.9	4.9	7.6	14.1	0.2
1800ª	1.1	3.3	4.3	24.7	0.2
1800ª	12.9	4.9	7.5	9.4	0.3
1800ª	0.5	2.0	4.0	27.4	0.1
1800ª	—	1.2	2.6	30.1	0.1
1900ª	4.0	7.4	18.0	35.7	0.2
1900ª	3.2	5.3	11.8	42.9	0.1
1900ª	12.3	7.8	23.7	16.0	1.3
1900ª	3.9	2.9	8.8	11.9	0.1
1900ª	7.1	3.3	11.4	7.9	0.2
1900ª	1.5	3.1	6.6	20.7	0.2
1900ª	8.9	10.5	15.2	28.2	0.5
1900ª	19.2	11.6	16.6	15.5	0.7
1900ª	3.8	7.8	11.0	41.0	0.3
1900ª	6.1	5.3	8.3	13.5	0.1
1900ª	1.4	3.7	4.7	23.7	0.1
1900ª	10.6	5.2	8.4	10.5	0.1
1900ª	0.3	1.3	2.7	29.7	—
1900ª	1.0	3.3	7.0	56.7	0.1
2000ª	3.2	5.3	13.4	46.1	0.2
2000ª	7.5	6.0	18.6	35.8	0.5
2000ª	0.8	0.3	33.9	32.5	0.5

ª See Table II.

TABLE IV. (*Continued*)

Base Carbon Conversion Range, Fraction	Rate Constant, $f_L k_T$, min^{-1}	
	Experimental	*Calculated*
0-0.277	0.0045	0.0034
0-0.130	0.0002	0.0006
0-0.143	0.0007	0.0009
0-0.159	0.0006	0.0006
0-0.228	0.0022	0.0044
0-0.571	0.0146	0.0093
0-0.106	0.0001	0.0001
0-0.258	0.0022	0.0033
0-0.176	0.0004	0.0002
0-0.228	0.0026	0.0036
0-0.703	0.0234	0.0233
0-0.426	0.0089	0.0099
0-0.745	0.0272	0.0400
0-0.550	0.0122	0.0122
0-0.576	0.0143	0.0128
0-0.619	0.0166	0.0161
0-0.543	0.0125	0.0096
0-0.817	0.0516	0.0481
0-0.736	0.02680	0.0216
0.43-0.871	0.104	0.128
0.43-0.663	0.0200	0.0169
0.43-0.659	0.0188	0.0274
0.43-0.661	0.0136	0.0172
0.43-0.556	0.0110	0.0117
0-0.452	0.0092	0.0095
0-0.842	0.0404	0.0315
0-0.952	0.0771	0.0588
0-0.697	0.0230	0.0217
0-0.623	0.0170	0.0175
0.43-0.933	0.0880	0.0880
0.43-0.979	0.236	0.248
0.43-0.776	0.043	0.0300
0.43-0.784	0.0395	0.0448
0.43-0.678	0.0299	0.0193
0.43-0.878	0.0736	0.0849
0.43-0.899	0.105	0.146
0.43-0.771	0.0431	0.0389
0.43-0.959	0.182	0.199
0.43-0.647	0.0111	0.0101
0.43-0.892	0.246	0.268
0.43-0.914	0.273	0.271
0.43-0.998	0.391	0.395
0.43-0.991	0.579	0.513
0.43-0.845	0.0820	0.0964
0.43-0.956	0.141	0.169
0.43-0.840	0.0638	0.0713
0.43-0.981	0.364	0.415
0.43-0.999	0.227	0.200
0.43-0.824	0.0721	0.0609
0.43-1.009	0.438	0.442
0.43-0.984	0.191	0.142
0.43-1.033	0.500	0.471
0.43-0.872	0.085	0.064
0.43-1.028	0.655	0.817
0.43-1.001	1.000	0.892
0.43-0.991	1.36	1.27
0.43-1.047	0.771	0.873
0.43-0.987	1.462	1.297

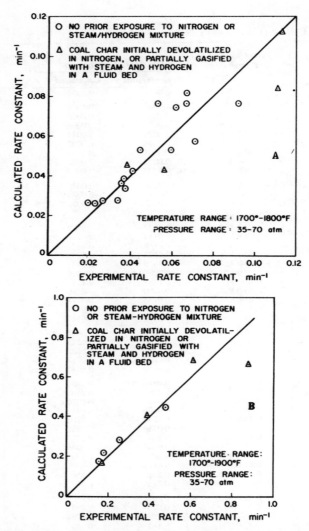

Figure 13 Comparison of experimental and calculated rate constants for low-rate gasification in steam–hydrogen mixture (thermobalance data).

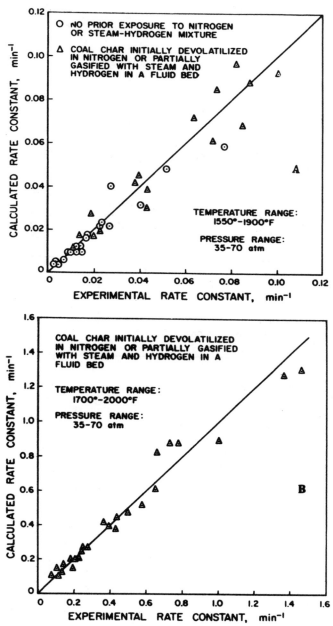

Figure 14 Comparison of experimental and calculated rate constants for low-rate gasification in synthesis gases (thermobalance data).

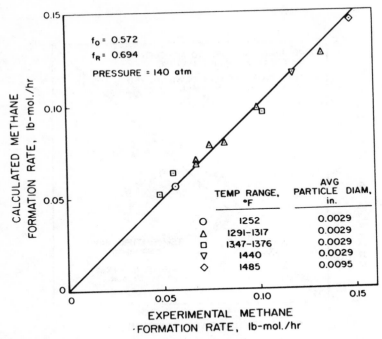

Figure 15 Experimental and calculated integral rates of methane formation for hydrogenation in a moving-bed reactor (IGT study).

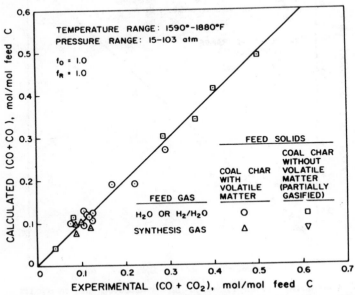

Figure 16 Experimental and calculated integral rates of carbon oxides formation for gasification in 2-, 4-, and 6-inch id fluid-bed reactors (IGT studies).

rates of carbon oxides and methane formation are compared in Figures 15-19 which show good agreement for a wide range of experimental conditions.

Frequently, the $P_{CH_4}/P_{H_2}^2$ ratio in product gases from integral fluid-bed systems for the gasification of coal or coal char with steam-hydrogen-containing gases is greater than the equilibrium constant for the graphite-hydrogen-methane system. This has often been interpreted as corresponding to a situation in which the coal or coal char has a thermodynamic activity greater than unity with respect to graphite. The models proposed in this paper offer two other explanations for this phenomenon. Rapid-rate methane formation occurs when coal or coal char containing volatile matter is used as a feed material. The methane yield resulting from this step is kinetically determined, independent of methane partial pressure. Under certain conditions then, values of $P_{CH_4}/P_{H_2}^2$ greater than that corresponding to the equilibrium for graphite-hydrogen-methane system

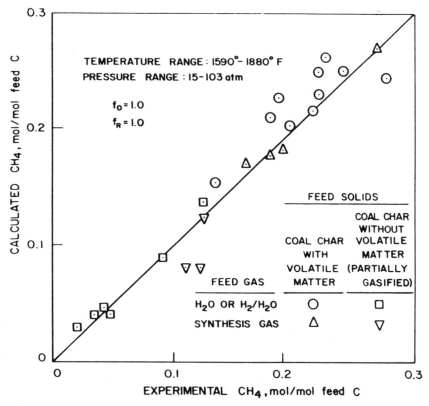

Figure 17 Experimental and calculated integral rates of methane formation for gasification in 2-, 4-, and 6-inch id fluid-bed reactors (IGT studies).

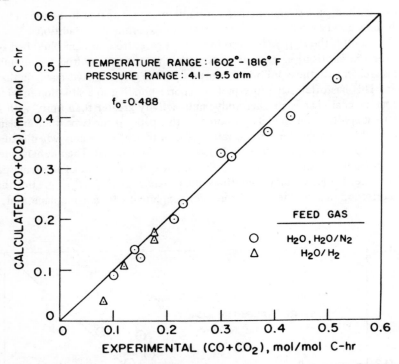

Figure 18 Experimental and calculated integral rates of carbon oxide formation for gasification in a fluid-bed reactor. Data from Ref. 18.

can result. Values of $P_{CH_4}/P_{H_2}^2$ greater than that corresponding to the equilibrium considered can also occur for low-rate gasification of coal char according to the model assumed in this study. This is illustrated in Figure 20 where the results shown were based on computations of gas yields in a hypothetical fluid-bed for char gasification with a pure steam feed gas using the correlations described above. The reason for the behavior illustrated is that at intermediate values of hydrogen partial pressure, the rate of Reaction III, which produces methane, is greater than the reverse rate of Reaction II in which methane is consumed when a potential for carbon deposition by this reaction exists. The partial pressure dependencies defined in the correlations developed are such, however, that at sufficiently high hydrogen partial pressures Reaction II dominates, and equilibrium for this reaction is approached. The qualitative trends in Figure 20 and even the magnitudes of these trends bear a striking resemblance to a similar plot given by Squires [19] to correlate the activities of coals and chars for equilibrium in the char–hydrogen–methane system with temperature and pressure.

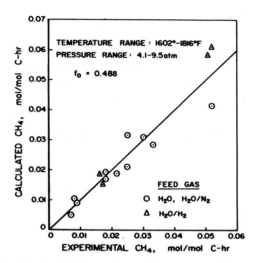

Figure 19 Experimental and calculated integral rates of methane formation for gasification in a fluid-bed reactor. Data from Ref. 18.

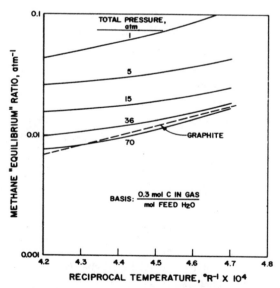

Figure 20 Calculated variations of the methane equilibrium ratio (P_{CH_2}/P_{H_2}) for gasification of carbon with steam in a backmixed reactor.

DEFINITIONS

A = nongasifiable matter in feed coal char (including ash and some sulfur), grams/gram feed coal char

C_b = base carbon in coal char at an intermediate gasification level, grams/gram feed coal char

C_b^0 = carbon in the fixed carbon fraction of the feed coal char as determined by a proximate analysis, grams/gram feed coal char

C_t^0 = total carbon in the feed coal char as determined by an ultimate analysis, grams/gram feed coal char

C_v = volatile carbon in feed coal, grams/gram feed coal char

f_0 = relative reactivity factor for low-rate gasification dependent on the particular carbonaceous solid

f_L = relative reactivity factor for low-rate gasification which depends on the particular carbonaceous solid and pretreatment temperature

f_R = relative reactivity factor for rapid-rate methane formation dependent on the particular carbonaceous solid

k_T = overall rate constant for low-rate gasification, min^{-1}

$k_2(T), k_3(T)$ = rate constants, min^{-1}

k_I, k_{II}, k_{III} = rate constants for Reactions I, II, and III, min^{-1}

$K_I^E, K_{II}^E, K_{III}^E$ = equilibrium constants for Reactions I, II, and III, considering carbon as graphite

N_{A_0} = concentration of species A_0 at any time, mole/mole base carbon

$N_{A_0}^0$ = initial concentration of species A_0, mole/mole base carbon

N_{A_*} = concentration of species A_* at any time, mole/mole base carbon

$P_{H_2}, P_{H_2O}, P_{CO}, P_{CO_2}, P_{CH_4}$ = partial pressures of H_2, H_2O, CO, CO_2, and CH_4, atm

t = time, min

T = reaction temperature, °R

T_0 = pretreatment temperature, °R

V = volatile matter in feed coal char (including moisture), grams/gram feed coal char

W_0 = weight of feed coal char, grams

ΔW = weight loss of coal char during gasification, grams

X = total base carbon conversion fraction

X_R = base carbon conversion fraction after reaction in rapid-rate methane formation stage is completed

α = kinetic parameter which depends on gas composition and pressure

$\beta = k_3(T)/k_2(T)$ ratio

γ = noncarbon matter evolved along with base carbon, grams/gram base carbon evolved

REFERENCES

1. Goring, G. E., et al., *Ind. Eng. Chem.* (1935) **45**, 2586.
2. Zielke, C. W., Gorin, E., *Ind. Eng. Chem.* (1955) **47**, 820.
3. Zielke, C. W., Gorin, E., *Ind. Eng. Chem.* (1957) **49**, 396.
4. Feldmann, H. F., Mima, J. A., Yavorsky, P. M., ADVAN. CHEM. SER. (1974) **131**, 108.
5. Zahradnik, R. I., Grace, R. J., ADVAN. CHEM. SER. (1974) **131**, 126.
6. Mosely, F., Patterson, D., *J. Inst. Fuel* (1965) **38**, 378.
7. Blackwood, J. D., McCarthy, D. J., *Aust. J. Chem.* (1966) **19**, 797.
8. Wen, C. Y., Huebler, J., *Ind. Eng. Chem., Process Des. Develop.* (1965) **1**, 147.
9. Eddinger, R. T., Freidman, L. D., Rau, A., *Fuel* (1966) **19**, 245.
10. Kimber, G. M., Gray, M. D., *J. Combust. Flame Inst.* (1967) **2**, 360.
11. Birch, T. J., Hall, K. R., Urie, R. W., *J. Inst. Fuel* (1960) **33**, 422.
12. Hiteshue, R. W., Friedman, S., Madden, R., *Bur. Mines Rep. Invest. No. 6376* (1964).
13. Pyrcioch, E. J., Feldkirchner, H. L., Tsaros, C. L., Johnson, J. L., Bair, W. G., Lee, B. S. Schora, F. C., Huebler, J., Linden, H. R., *IGT Res. Bull. No. 39*, Chicago, Nov. 1972.
14. Blackwood, J. D., *Aust. J. Chem.* (1959) **12**, 14.
15. Goring, G. E., et al., *Ind. Eng. Chem.* (1952) **44**, 1051.
16. Blackwood, J. D., McGrory, F., *Aust. J. Chem.* (1958) **11**, 16.
17. Curran, G., Gorin, E., *U.S. Off. Coal Res. R&D Rept. No. 16, Interim Rept. No. 3, Book 2*, Government Printing Office, Washington, D.C., 1960.
18. May, W. G., Mueller, R. H., Sweetser, S. B., *Ind. Eng. Chem.* (1958) **50**, 1289.
19. Squires, A. M., *Trans. Inst. Chem. Eng.* (1961) **39**, 3, 10, 16.

RECEIVED May 25, 1973. Work sponsored by the American Gas Association, U.S. Department of the Interior, Office of Coal Research and the Fuel Gas Associates.

Index

*NOTE Letters "t" and "f" following entry numbers refer to tables and figures, respectively.

Activation-energy distribution, 92-95, 92f
Active site function, 110f
Adsorption isotherms, carbon dioxide, 51; see also Carbon dioxide, adsorption; Surface areas
Aggregative (gas-fluidized) beds, 20
Aliphatic hydrocarbon side chains, 64
Alkali catalysts, 176, 274; see also Catalysts; Coal-gasification systems, catalysts
Alkaline earth metals, 272
Ammonium molybdate, 267, 268
Amorphous carbons, 3, 4, 19
Asphaltenes, 267

Backmixed reactor, 313f
Base carbon, 72, 288
    conversion fractions, 44, 51, 240, 249f, 250f, 255f, 256f, 257f, 259f, 291f, 292f, 293f
    gasification
        low-rate, rapid-rate methane, 290
        formation, 290
Benzene yields, 221f
Benzypyran structures, 101
BET equation, 51, 238
Bituminous coal, 78, 166, 273
    carbon conversion in, 57
    devolatilization of, 98
    gaseous diffusion of, 81
    gasification, 74, 199
        fixed bed, 60
        fluidized bed, 60
        in hydrogen, 89f, 90f, 188
        rate of, 131
    German, high-volatile, 106
    high-volatile, 151, 180, 202
        air-pretreated, 74f, 75f

318 INDEX

in carbon dioxide, 130
in hydrogen and helium, 76f
hydrogasification, 43
initial stage coal-gasification reactions, 62t, 63t
noncaking, high-volatile, 66
particles, fixed bed, 81
pore volumes, 41-43; *see also* Surface areas, pore volumes
Pittsburgh seam, 101
raw
  distribution of volatiles in, 19t
  hydrogasification in, 233
  reaction-heat measurements of, 15
  surface-area variations of, 57f
  weight-loss fraction for, 78-79
Bituminous coal char, 104
  air-pretreated, hydrogen gasification in, 102
  British noncaking, high-volatile, 106
  gasification
    in carbon dioxide, 123
    in carbon monoxide-carbon dioxide, 122
    kinetics, 283-315
  high-volatile, 112
  high-volatile (Pittsburgh seam), 128
  high-volatile, pretreatment temperatures of, 109
  low-temperature, 3, 66
    gasification, 65
  low-volatile, 300
  reaction-heat measurements of, 15
Brazilian metallurgical coal, 252
Brown coal, 64
  gasification of, 59, 64
  hydrogenation, fluid bed, 292
  integral fluidized bed gasification, 3
  particle size, 162
Brown-coal char, 104
  gasification of, 103, 112
  preparation temperature, 107
Bubble cloud, 22, 23f
Bubble phase, 21, 25
Bubble velocity, 22-23
Bubbling fluidized beds, 21-22, 21f, 23f, 24-25, 27t

Calorimetry, 13
Carbon
  gasification of, 145, 155
    in pure steam, 152f
    in steam-hydrogen mixtures, 138

rates, 98, 133, 154, 303
  comparison of, 140f
residual oxygen in, 101
surface adsorption, 52; *see also* surface areas
Carbonaceous solids, effects of catalysts in gasification of, 271
Carbon burn-off, 44
Carbon-carbon dioxide reaction, 120, 132, 134, 136, 141, 142, 148
  effect of hydrogen concentration on, 136
  isotopic techniques in, 120-121
Carbon conversion
  deficiency, 137
  excess, 72, 137
  levels, 44f, 45t, 46, 98, 137, 151, 168
Carbon crystallite, 101
Carbon dioxide
  adsorption, 53
  adsorption isotherms, 51, 52
  chemisorbed, 121
  gasification rates, 118, 132
  yields, 82, 215f
Carbon dioxide-carbon monoxide mixtures
  equilibrium constants in, 144t
  gasification in, 143
  steady-state gasification in, 137
  with steam and hydrogen, 142
Carbon-hydrogen
  reaction, 132, 134, 137
  systems, 120, 137
Carbon-hydrogen-oxygen system, 6-9, 7f, 8t, 137
Carbon hydrogenation, 17
Carbon monoxide
  carbon-steam reaction, 141
  concentration effects on gasification, 135f, 136f
  formation rates, 150f, 154, 155, 159, 159f
  gasification temperature increase, 135
  partial pressure increase, 134
  yield, 215f
Carbon oxides
  formation, kinetics of, 138, 152
  methanation, 61
Carbon-oxygen
  complex, 119, 158
  system, 137
Carbon-solid phase, 8
Carbonization product yields, estimation of, 19
Carvao metallurgical coal char, 252f

INDEX 319

Catalysts, 164, 261-281; *see also* Alkali catalysts, Coal-gasification systems, catalysts in
  alkali, 167
  in char gasification, 163
  physically mixed, 167
  poisoning by sulfur, 168, 277
  Raney nickel, 167
  recovery of, 276, 277
  sulfur-resistant, 168
  thermal efficiency variations of, 276f
Catalyst activity, loss of, 276
Catalyst impregnation, 165, 272
Catalytic reactor systems, 28
C_3^+ gaseous hydrocarbons, 182
  evolution of, 82
  gas-phase hydrogenation, 84
  hydrogenation, 188
  secondary hydrogenation, 89, 186
  yields, 187f
Char, 96
  air-pretreated hvab Pittsburgh, 287
  conversion rate, 129
  gasified in carbon dioxide, specific rates, 133t
  gasified in hydrogen, 49, 104f, 132
    rapid-rate, 114
    reactivity variations, 111, 132
    steady-state rate, 115-117
  gasified in hydrogen-methane mixtures, kinetics of, 106
  gasified in pure steam, 151
  gasified in steam-hydrogen mixtures, 49, 150, 157, 159f
  inherent reactivity, 112
  low-rate gasification, 312
  potential volatile matter, 10
  pore structure of, 161-162
  reactivity factors, 242
  residue, 71
  subbituminous, 129
Char-carbon dioxide-carbon monoxide system, 98, 117-124
Char conversion rate, 129
Char formation, reactive-gas atmosphere, 58
Char gasification, 263
  bituminous coal, kinetics of, 283
  catalysis, 165, 172
    rates of, 114, 251
      differential, 148
      transient effects, 113, 115

  use of, 163
  film-diffusion effects, 161
  internal particle diffusion, 161
  kinetics of, 84, 96, 272, 283, 288, 296
  low-rate, 296
  methane production in, 271; *see also* methane
  nitrogen carbonization, 58
  particle-size effects, 161
  reactivity factors in, 112-113, 113f
  reactions, 96, 237-260
  reaction systems, 98
  transient effects in, 113, 115
Char-gasification stage, experimental studies in, 271
Char-hydrogen gasification rates, pretreatment temperatures, 132, 150
Char-hydrogen reaction, 105t
Char-hydrogen-methane system, 90, 98, 99, 106, 148, 312
Char-hydrogen-steam-carbon monoxide-carbon dioxide-methane system, 98
Char particles, gasified in carbon dioxide, 47
Char pore volumes, variations during gasification, 253; *see also* Surface areas, pore volumes
Char-preparation temperatures, 106, 107f, 110f
Char residues, base-carbon conversions, 48; *see also* Base carbon, conversions
Char shrinkage during gasification, 51
Char-stage reactions, 263, 264
Char-synthesis gas system, 137
Char surface areas, internal variations during gasification, 247; *see also* Surface areas, char
Char weight-loss fraction, 44
Chemisorbed carbon monoxide, 121
Chemisorbed hydrogen, 144
Chromene structures, 101
Circular pores, 39
Coal, direct hydrogenation of, 78
  initial gasification stages, 71
  low-rank initial gasification stages, 82
  nonagglomerating, 81
  potential volatile matter in, 10
  shape factors of, 29t
Coal-carbon conversion, 212f
Coal composition, 36
  analysis of, 183t, 248t
  carbon content, 132

effects on gasification rate in carbon dioxide, 133f
Coal density, 39
Coal-devolatilization reactions, 4; see also Devolatilization
Coal gasification, 1, 237-281
　bituminous, 199
　　kinetics, 179
　component conversion, 184f
　enhanced yields, 81
　fluidized bed advantages, 22
　fluidized bed disadvantages, 22
　in hydrogen, 184f
　initial, effects of diffusion inhibition, 81
　low-rate, 299f, 300f, 301f, 302f, 305f, 308f, 309f
　methane formation in initial stages, 69
　thermodynamic characteristics of, 2, 4
　use of catalysts in, 261; see also Catalysts, alkali catalysts; coal-gasification systems, catalysts in
coal-gasification fluidized-bed reactors, 24
coal-gasification systems
　carbonization heat in, 13-15, 14f, 15f
　catalysts in, 163-169, 261-281, 274, 277
　catalytic effects of, 261
　continuous, 269
　equilibrium characteristics of, 2-3
　experimental studies of, 266, 271
　flow sheet in, 262f
　general characteristics of, 262
　heat effects, computation of, 19
　hydrogen-helium in, 78
　physical processes of, 20
　reaction-heat effects in, 12
Coal gasifiers, design, 59
Coal hydrogasification, 92
Coal hydrogenation, 267
　catalysts, 268
　dilute-phase flow reactor, 94
　direct, 186
　kinetics, 186, 189
　direct, rapid-rate, 84
Coal-hydrogen evolution, 83, 87, 92, 189, 190f, 191f, 197
　conversion, 197f, 214f, 222f
　kinetics, 85, 88, 91
　carbon ratio, 228f
　secondary stage, 69
Coal-hydrogen reaction, 79
Coal lithotypes, 55

Coal oxygen, 82, 270f
Coal-oxygen conversion, 213f
Coal pyrolysis, 60; see also Pyrolysis
Coal-stage reactions, 263
Coal types, 242t
Cobalt molybdate, 165, 272
Coconut char, 104
　carbon solids in, 128
　carbon-steam reaction in, 141
　gasification, 120, 131
　gasification kinetics, 152
　steam-hydrogen mixtures, 152
Coke, high-temperature, 37
Coke, metallurgical, gasification of, 142
Coking, reaction-heat effects, 12
Composite rate constant, 128f
Crystallite reorientation, 51

Decomposition-polymerization process, 68
Dense solids concentration, 24
Densities, solids, coals, and coal chars, 37-39, 38f
Depolymerization reaction, 60
Devolatilization, 18, 60, 64, 65, 179, 283, 288, 305
　first-stage reaction, 83
　fixed beds, bituminous coals, 98
　incomplete, 14
　initial stage reaction, 64
　inert-gas atmosphere, 59, 60
　primary, 60f, 61, 82
　processes, 82
　secondary, 61, 61f, 87, 182
　temperature effect on internal surface, 111t
Differential scanning calorimetry (DSC), 13
Differential thermal analysis (DTA), 14
Diffusion inhibition, 81
Dilute phase flow reactor, coal hydrogenation, 94
Dilute-phase systems, 205
　solid-phase transport, 43, 64
　transport reactor, 66, 70, 82, 88, 96, 269
Dilute solids concentration, 24
Disco char, gasification of, 103, 147, 150, 284, 303, 304
　reactivity, 300
　in steam-hydrogen mixture, 150f, 151f
Dissociation steam chemisorption, 14
Dolomitic limestone, 274
DSC, 14
Dubinin-Polanyi equation, 51, 52

INDEX

Durain, 55

Elutriation, 31
Emulsion phase, 22
  dense-solids concentration, 25
  gas-flow patterns, 25
  high solids density, 21
  lean-solids concentration, 25
Endothermic heats of reaction, 14
Endothermic pyrolysis, 16
Ethane, 83, 84

Feeder pores, macro and transitional, 37; *see also* Surface areas, pore structure
Feed solids, ash content, 13
Fixed-bed flowing gas systems, 96, 267
Fixed beds, 96, 205
  devolitalization of bituminous coals in, 98
Fixed beds-flowing gas, 64
Fluid bed systems, 292f, 305, 310f, 311f, 312f, 313f
Fluidized beds, 96, 205
  bubbling, *see* Bubbling fluidized beds
  catalytic decomposition of ozone, two-dimensional, 28
  diagram of, 21f
  expansion, 31, 33t, 34
  experimental studies, 28, 47f, 62t, 63t
  general properties, 28
  initial stage reactions in, 62t, 63t, 64
  minimum fluidization velocity, 29-31, 32t
  particle-shape factors, 28
  solids circulation, 34-35
  two-phase models (gas-solid contracting configurations), 24-25, 24f, 26t
  voidage at minimum fluidization, 28
Fluidized-bed reactor, 55
  designing, 25
  scaling up, 25
Fluidized-bed systems, physical processes, 3, 20
Fluidized-bed technology, 23
Fluidized coal bed, 64
Free-fall systems, 43, 64

Gas-adsorption isotherms, 51
Gas composition, gasification of coal chars, 4
  equilibrium calculations in, 4
Gaseous hydrocarbons, 61
  light, 78
Gaseous hydrogen, 84
  production of, 80

Gas-fluidized beds, characteristics of, 20
Gasification
  initial stages, kinetics of methane and ethane formation, 223
  low-rate, 283, 288
  low-rate, kinetics of, 297
  steady-state, 102
  coal char, 288
  low-rank coals, 180, 182
Gasification rates
  comparison of specific, 131t
  effects of pressure on, 132t
  in pure carbon dioxide, 129f
  total pressure, 122
Gasification reactions, kinetics of, 59
Gasifications systems, physics, 20
Gasifications thermodynamics, theoretical evaluation of, 2
Gasified chars, internal surface area, 51
Gas-solid contacting mode, 143
Graphite-carbondioxide-carbon monoxide system, 5, 5f, 6f
Graphite gasification, 139
  hydrogen, 3
  hydrogen-methane system, 6, 106, 311

Helical-coiled transport reactor, 206
Helium
  mixed with hydrogen, 78, 82
  pure, 78, 82
Helium displacement, 37
High-temperature cokes, *see* Coke, high temperature
High-vitrinite material, 53
High volatile bituminous coal, *see* Bituminous coal, high-volatile; Bituminous coal char, high-volatile
Hvab coal char, 122
Hvab coals, carbonization of, 55
Hvab Pittsburgh coal, hydrogenation of, fixed bed, 292
Hvab Pittsburgh coal char, air-pretreated, 286t, 287
Hydrated lime, 274
Hydrocarbon product species, 216
  heavy-hydrocarbon yields, 219f
  liquid, formation of, 268f
  unknown-hydrocarbon yields, 220f
Hydrochloric acid, 244
Hydrogasification, 66, 87
Hydrogasification, raw, bituminous coals, 233

322 INDEX

Hydrogen, 82
  gasification, 80
    inhibition of rates, 139
    relative-reactivity-factor correlation, 150
  mixture with helium, 78, 82
  pure, 78
  partial pressure, 229f
Hydrogenation, 66, 67, 69, 100
  C_3^+ hydrocarbons, 188
    heterogeneous, 69
    volatile matter, 16, 59, 60
  heat reactions, 16f, 18f
  process, 80
  of volatiles, methane yields in, 71
Hydrogenation products, 61
Hydrogenation reactions, 65
Hydrogen atoms, 140
  surface, formation, 101
Hydrogen chemisorption process, 100
Hydrogen evolution, 61
Hydrogen gasification, air-pretreated bituminous coal char, 102
Hydrogen-helium mixtures, 78, 82
Hydrogen-methane mixtures, equilibrium measurements, 3
Hydrogen molecule, 100
Hydrogenolysis, volatile matter, 59
Hydroxyl radicals, 140

Illinois No. 6 high-volatile bituminous coal, gasification of, 88
Illinois seam coals, 130
Inert-gas atmosphere devolatilization, 59, 60
Institute of Gas Technology's HYGAS Process, 264
Integral fluid-bed systems, product gases, 311
Integral fluidized-bed gasification, brown coal, 3
Integral steam-oxygen gasification, coals and coal chars, 3
Ireland mine coal char, 289-290
  air-pretreated, 291, 300, 302, 304, 305
  rate constants for, 303t
    in hydrogen, 303t
    in steam-hydrogen mixtures, 304t
    in synthesis gas, 306t
Isotopes, 120

Japanese vitrains, 53
  total pore volumes, 40; *see also* Surface areas, pore volumes

Kinetics, first order, 91
Kinetic models, 287
Kinetic parameters, 70, 70t, 73, 78, 91-92, 99, 126, 128, 130, 135, 138, 139, 143, 144, 147, 148, 149, 154, 156, 158, 189, 194, 195f, 196, 199f, 229f
  comparison of, 112t, 124t, 125t
  values of, 155t

Lewis acidity, 269
Lignite, 57, 66, 273
  acid-treated, 57
  gasification, 65, 71, 72, 74, 132
  gasification kinetics, 82
  in steam-hydrogen mixture, 46-47
  Montana, 88
  North Dakota, 88
  pore volume distribution, 49; *see also* Surface area, pore variation
  untreated, 58
Lignite ash, 274
Lignite char
  gasification, 56
  pretreated in nitrogen, 110
  Renner Cove, 157, 160
Lignite-char-gasification reactivities, effects of exchangeable cation, 166
Lignite-gasification reactivities, 165, 243, 245f, 272, 273
  catalytic effects of exchangeable cations, 165
Low-vitrinite material, 55

Macerals, 36
Maceral concentrates: exinite, fusinite, vitrinite, 37
Macropores, 36, 49, 51
Macropore volume, 255, 256
Mass spectrometer, 285-287
Mass-transfer inhibition, 82
Mercury porosometer, 257
Mercury porosometry, 49
  measurements in, 41
Metal halides, 268, 269
Metallurgical coke, kinetic results, conversion system, 14
Metastable intermediate product, 60
Methanation requirements, 265t
Methane, 61, 71, 84, 297
  direct formation, 265
  indirect formation, 265
Methane formation, 61, 65, 68, 154, 277

INDEX 323

rapid-rate, 87, 90, 91, 93, 106, 223-234, 283, 287, 288, 291f, 292f, 293f, 295f, 305
  base carbon conversion, 294
  gasification of base carbon, 290, 292
  weight loss, 290
Methane formation kinetics, rapid-rate, 88
Methane formation rates, 68-69, 153, 156-157
  comparison of, 151f, 156f
Methane formation reactions, 60, 160
  rapid-rate, 87
Methane-hydrogen mixtures, 88
Methane mole fractions, 5f, 6, 9
Methane production, char gasification, 138, 271
Methane-plus-ethane formation, 186, 191, 216-217
  coal thermal decomposition, 188
  initial stage kinetics, 186
  ratio, 218f
Methane-plus-ethane formation kinetics, initial stages of gasification, 223
  rapid-rate, 224
Methane-plus-ethane yields, 85, 85f, 87, 89, 89f, 90f, 182, 185, 187f, 190f, 197, 198f, 200f, 201f, 217, 217f, 218f, 222f, 231f, 232, 232f, 233f, 234
Methane reaction, rapid-rate, 84, 90, 115
Methane yields, 2, 8, 66, 67, 78, 94f, 274, 275
Micropores, 36, 49
Micropores surface area, 52
Micropore volume, 52, 53, 255, 256
Microwave fields, 101
Mineral constituents, distribution characteristics, 36
Mineral heterogeneities, 53
Molecular dissociation, 101
Molybdic oxide, volatile, 268
Montana lignite, 85, 88, 223
  acid-treated, 246
  base-carbon conversion fraction for, 249f
  feed composition, 207t, 211, 211t
  gasification, 55, 185, 186, 233
    in hydrogen or helium, 187f, 191, 205-235
    kinetics, 182
  initial stage reaction, 205
  pyrolysis of, in helium, 187f
Montana lignite char reactivities, 227, 247f
  effects of exchangeable calcium and sodium concentrations, 245, 246t
Montana subbituminous coal, 88
  gasification kinetics, 182

Moving-bed systems, 305, 310f

New Mexico subbituminous coal, 267
Nickel catalyst, 276
  commercial, 275
Nitrogen, capillary condensation, 51
Nitrogen adsorption isotherms, 41, 49
Nitrogen isotherms, 51
Nonagglomerating coal, 81
  catalyzed gasification, 82
North Dakota lignite, 88, 300
  gasification kinetics, 182
Nuclear reactors, 164
  gas-cooled, 271

Oil, distillable, 267
Oil formation, 268
Orifice meter, 285
Oxygen-containing functional groups, 64
Oxygen distribution, evolved, 216f
Oxygen-exchange reaction, 119, 141, 143
Oxygen-exchange equilibrium constant, 127f, 143
  comparison of in carbon dioxide-carbon monoxide reaction, 144t

Particle shrinkage, 46
Particle-size effects on gasification, 161
Pittsburgh coal char, air-pretreated hvab, 287
Pittsburgh seam bituminous coal, 101
Pittsburgh seam coals, 130, 147
Polymerization, 13
Pore structure, characteristics of, 35-37
Pore surface, see Surface areas
Potassium, 276
Precarbonized chars, 43-44
Primary volatile matter, 10
Pseudoequilibrium constants, 2
Pyroheat capacity, 16
Pyrolysis, 69, 106, 264
  coal, 60
  endothermic, 16
  methane yields, 71, 72f
  thermal, 287
Pyrolysis hydrocarbons, 17
Pyrolysis model, 60
Pyrolysis processes, 65, 263
  primary, 82
Pyrolysis reactions, 66, 223
Pyrolysis yields, 89

INDEX

Raney nickel catalyst, 167, 274, 275
Rapid-rate methane formation, 96
Raw bituminous coals, hydrogasification, 233
Reactor coil data, 208t
Reactivity factors, 113f, 132, 133f, 237-260, 243f, 253f
Reactivity ratios, 245t
Redox reactions, 101
Renner Cove lignite char, 60, 157

Seam coals, Illinois, 130
Seam coals, Pittsburgh, 130
Secondary gas-phase reactions, 13
Secondary volatile matter, 10
Semichar, 60, 61, 83, 84, 88
Shape factors of coal, 29t
Shrinking-core phenomenon, 258
Slag formation, 274
Solids
   convective movement of, 22-23
   density, emulsion phase and bubble phase, 22
   density of coals and coal chars, 37-39, 38f
   elutriation, 22
Steam-carbon reaction, kinetics, 139
Steam-hydrogen gasification, 3
Steam-oxygen gasification, 3
Subbituminous A coal, 57, 251f
Subbituminous char, 129
   gasification rate, 111, 131
   gasification systems, thermodynamic characteristics of, 19
   kinetics, 82
Sulfur poisoning, 274
Surface areas
   carbon dioxide, 52
   char pore volumes, 49, 49f, 50f
   char structure, 36
   feeder pores, 37
   internal carbonized char, 51

   internal, devolatilization temperature effects on, 111t
   nitrogen, 52
   pore structure, 35-36, 162
   pore volumes, 39-51, 41f, 42f, 43t, 44f
   pore variations, 255
   transitional pores, 36, 49, 51
   transitional pore structures, 53
Synthane process, 273, 274

Thermal deactivation, 86
Thermal decomposition, 13
Thermal decomposition-polymerization process, 68
Thermal pyrolysis, 288
Thermobalance, 284
   high pressure, 286f
Thermobalance reactor, 285f
Through-flow gas, 23
Transport reactor, continuous dilute phase, 206
   dilute-solids-phase, 269
   helical-coiled, 206

Volatile carbon, 288
Volatile matter, hydrogenation, 60
Volatiles, 77
   distribution of, 19
   nonreactive species, 75, 77
   reactive species, 75, 76, 77
      hydrogenation, 78
      stabilization, 79

Water-gas shift reaction, 147, 148, 152
Wet-test meter, 285
Wood char, 104
   effect of preparation temperature on, 134f
   gasification rates in carbon dioxide, 132

Zero solids concentration, 24

RETURN TO ➡ CHEMISTRY LIBRARY
100 Hildebrand Hall   642-3753

LOAN PERIOD 1	2	3
~~7 DAYS~~	1 MONTH	
4	5	6

ALL BOOKS MAY BE RECALLED AFTER 7 DAYS
Renewable by telephone

DUE AS STAMPED BELOW

~~DEC 6 '80~~	~~SEP 10 '82~~	MAR 01
~~MAR 21 '81~~	~~DEC 4 '82~~	
~~MAR 21 '81~~	~~MAR 1983~~	
~~MAR 21 '81~~	~~NOV 2 1983~~	
~~MAR 23 1981~~	NOV 17 1989	
~~JUN 13 '81~~	JUN 1 1993	
~~SEP 12 '81~~	6/7/93	
DEC 12 '81	APR 15 1996	
	MAR 22 REC'D	
~~MAR 27 '82~~	MAY 03 1998	
~~JUN 19~~	MAR 22 '04	
~~JUN 19 '82~~	OCT 07 2006	

UNIVERSITY OF CALIFORNIA, BERKELEY
FORM NO. DD5, 3m, 3/80   BERKELEY, CA 94720